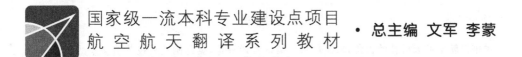

国家级一流本科专业建设点项目
航空航天翻译系列教材 · 总主编 文军 李蒙

航空航天英汉翻译教程

English – Chinese Translation Course for Aeronautics and Astronautics

主 编 李 蒙 彭 莹

北京航空航天大学出版社

内 容 简 介

本书以航空航天科技的视角,来诠释科技英语翻译理论,据此阐述科技英语文本的翻译策略和实用技巧,以提高航空航天专业人员的科技翻译水平,服务我国航空航天技术的国际交流和发展。书中依据从点到面、由易到难的原则,首先介绍航空航天科技英语的文体特征、语言特点及文本翻译标准;接着讲解词汇翻译策略,包括词义选择、词类转换、词义引申、增译和省译等;随后以句法翻译策略为主,重点阐释名词化结构、被动语态、无人称句、非限定动词、从句和长难句等难点;最后解决篇章翻译过程中的修辞和逻辑问题。本书不仅每章均附有针对性练习,还精选 23 篇文章作为翻译练习的素材,包括航空航天科普作品、说明书、科技报告和新闻报道。每篇文章都给出了必要的注释和参考译文,手把手指导读者进行翻译实践。

本书可作为航空航天类院校英语专业必修课教材,也可作为其他理工科院校非英语专业硕士生、博士生选修课教材,并可供科研工作者和翻译爱好者自学使用。

图书在版编目(CIP)数据

航空航天英汉翻译教程 / 李蒙,彭莹主编. -- 北京:
北京航空航天大学出版社,2023.10
ISBN 978 - 7 - 5124 - 4204 - 7

Ⅰ. ①航… Ⅱ. ①李… ②彭… Ⅲ. ①航空工程-名词术语-英语-翻译-教材②航天工程-名词术语-英语-翻译-教材 Ⅳ. ①V

中国国家版本馆 CIP 数据核字(2023)第 201452 号

航空航天英汉翻译教程
主 编 李 蒙 彭 莹
策划编辑 赵延永 蔡 喆
责任编辑 蔡 喆
*
北京航空航天大学出版社出版发行

北京市海淀区学院路 37 号(邮编 100191) http://www.buaapress.com.cn
发行部电话:(010)82317024 传真:(010)82328026
读者信箱:goodtextbook@126.com 邮购电话:(010)82316936
北京凌奇印刷有限责任公司印装 各地书店经销
*
开本:710×1 000 1/16 印张:15.75 字数:336 千字
2023 年 10 月第 1 版 2023 年 10 月第 1 次印刷
ISBN 978 - 7 - 5124 - 4204 - 7 定价:49.00 元

编 委 会

科学技术的发展离不开交流与合作,航空航天的发展也不例外。在中国航空航天发展史上,这种交流与合作很大程度上与翻译相关,概括起来,大致经历了两个阶段:早期的翻译引进,后期的翻译引进与输出并举。最早与航空航天有关的翻译引进活动始于 1903 年到 1907 年期间中国掀起的"凡尔纳热",其中的航空航天科幻小说翻译包括鲁迅的《月界旅行》、商务印书馆出版的《环游月球》以及谢祺的《飞行记》等。1910 年高鲁翻译出版了《空中航行术》,这是中国航空航天科技书籍和资料汉译的开端。而在译出方面,随着我国航空航天事业的飞速发展,近些年的科技新闻、政府白皮书等都有大量航空航天方面的信息对外发布,及时而系统地向全世界展现了中国在此领域的发展现状和巨大成就。

总体而言,航空航天的领域宽广,翻译多种多样。从翻译的主题看,航空航天话语以科技语言为主,其一般特点有七个方面:无人称、语气正式、陈述客观准确、语言规范、文体质朴、逻辑性强和专业术语性强[1],与之相关的科技论文等翻译是航空航天翻译的主体。此外,航空航天话语中还包括与商务活动相关的商贸翻译(如合同、谈判等)、与航空航天新闻活动相关的新闻翻译(如新闻发布会、各种媒体的相关新闻报道等)、与航空航天文学相关的文学翻译(如航空类小说、航天类科幻小说等)、与航空航天影视活动相关的影视翻译(如纪录片、科幻电影)等;从翻译活动的方式看,航空航天翻译包括了笔译、视译、交替传译、同声传译、机器翻译+译后编辑等几乎所有翻译方式。

航空航天翻译主题和体裁的多样性及翻译方式的全面性,对翻译人才的培养提出了新的、更高的要求。为此,我们特设计和编写了这套"航空航天翻译系列教材",其特色主要体现在以下几个方面:

1. "入主流"与"显特色"并举。"入主流"主要指各种教材的设计都体现了翻译这一核心要素,其内容选择都以"怎么翻译"为焦点;"显特色"则体现在教材内容的选

---

[1]　冯志杰.汉英科技翻译指要[M].北京:中国对外翻译出版公司,1998:6-7.

择上,无论是例句还是练习,都择选了与航空航天密切相关的语料,力求解决航空航天翻译中的实际问题。

2. 理论与实践并重。在教材设计上,突显理论融于实践的理念,对理论不做大篇幅的阐释,而将翻译策略、翻译方法等融于对例句和语篇的解释之中,而这些例句和语篇都选自真实的航空航天语料,以着力提升学生的翻译实践能力。

3. 阐释与练习并立。对各种翻译现象的解释与阐释在教材中必不可少,是教材的主干;与此同时,各教材采用按"节"的方式设置练习,其用意在于着力加强练习与教材正文的关联性,以方便学生的学习和操练。

本系列教材可以作为翻译专业、英语专业和大学英语相关课程的课堂教学材料,也可供对航空航天翻译感兴趣的读者使用。

迄今,本系列教材已规划了英汉翻译、汉英翻译、口译、影视翻译等教材;今后,我们还可增加与航空航天翻译相关的品种,如航空航天文学翻译、航空航天新闻翻译、商贸翻译、航空航天同声传译等教材。

为使本系列教材的编写更具广泛性和权威性,我们组建了高水平的编委会。编委会委员有北京航空航天大学文军、李蒙、梁茂成,北京理工大学李京廉,重庆大学彭静,大连理工大学秦明利,哈尔滨工业大学李雪,哈尔滨工程大学朱殿勇,华中科技大学许明武,南京航空航天大学范祥涛,南京理工大学赵雪琴,西北工业大学孔杰,西安航空学院张化丽和中国民航大学张艳玲等专家学者。

本系列教材的编写是一种尝试,希望得到业内专家学者、学生和其他读者的反馈和意见,以使教材更臻完善。

文军　李蒙
2023 年 3 月于北京

航空航天科技是科技进步和创新的重要领域,航空航天科技成就是国家科技水平和科技能力的重要标志。为了更好地服务于国家的航空航天战略,培养更多合格的航空航天领域科技翻译人才,我们编写了这本《航空航天英汉翻译教程》。

本教材的编写有以下几个特点:

1. **突出科技翻译的实践性**。在编排方面,对译学理论和翻译基本技巧的讲解力求简明扼要,对翻译练习材料精挑细选,尽可能在篇幅有限的情况下,为学生提供全面、系统、有针对性的翻译练习。全书分为理论篇和实践篇。理论篇中的练习以词组、单句和段落为主,在书末附有参考答案;实践篇选取了23篇科技篇章,采取原文+参考译文+注释的编排方式,方便学生对照练习。针对原文中的理解难点和翻译要点,从背景知识、语法结构、翻译技巧和汉语表达等各方面给予学生必要的指导。这些练习有助于学生理论联系实践,切实提高科技翻译实践技能。

2. **翻译材料内容新、体裁广**。在选择翻译材料时,注重时效性、知识性、趣味性和典型性。在实践篇科技新闻报道部分,选用近两年国外报刊与航空航天科技相关的最新报道,将翻译材料和最新热点时讯相结合,激发学生的学习热情。在科普文章部分,精挑细选的文本有助于学生拓展科技视野,增加科技知识储备,提升翻译能力。

3. **注重文体分析意识的培养**。文体与翻译关系紧密。在不同语境选择最有效的表达方式一直是文体学强调的重点,而译文语言的得体性也是翻译学研究的重点之一。为此本书第一章详细讲述了科技文本不同的文体特征,而在翻译转换中,充分强调译文文体的适应性。根据文本题材和体裁,将实践篇文章细分为航空航天科普作品、航空航天科技说明书、航空航天科技报告和航空航天科技新闻报道四大部分,并分别对其文体特征、语言特点和翻译要点进行介绍,帮助学生更有效地进行翻译实践。

4. **强调科技文本中的逻辑和修辞问题**。科技文本是科学与文学艺术的结合体,其科学性主要体现在逻辑上的高度严谨性,艺术性主要通过各种修辞手段的巧妙应用得以实现。翻译过程要求科学性和艺术性的统一。好的科技翻译要讲究"理"通

"词"顺,内外兼顾,既要注重客观准确、条理清晰、逻辑严密地呈现原文的科学信息,又要兼顾译文表达的地道自然,力求生动形象地反映客观事实。为此,专门设计了第4章"航空航天科技英语翻译中的修辞与逻辑",引导学生从逻辑和修辞视角去审视翻译,进行翻译实践活动。

本教程的编者多为高校教师,具有丰富的翻译教学和实践经验。全书由李蒙负责总体设计,制定编写原则和大纲,彭莹负责具体指导、统筹调度及统稿。

本教程承蒙北京航空航天大学外国语学院文军教授的指导,北京航空航天大学外国语学院翻译专业研究生杨腾腾、范美玉、梁晓纯、杨芃、刘静涵、姚紫芃同学也做了大量的工作,在此深表感谢。

在本书的编写过程中,参阅了许多专家和学者的研究成果,在此表示衷心的感谢。由于水平有限,教材中难免有疏漏之处,欢迎广大读者批评指正。

李 蒙　彭 莹
2023 年 3 月

目　　录

理 论 篇

实 践 篇

理　论　篇

第1章　航空航天科技英语与翻译概述

　　航空航天是人类开发大气层和宇宙空间时发生的活动的总称,与之有关的科学技术是 20 世纪兴起的一门跨学科的现代科学技术,高度综合了力学、热力学、材料学、电子技术、自动化技术、计算机技术和医学等基础和应用科学的最新成就以及工程技术的最新成果,已经形成了完整的体系。航空航天科技在推动原始创新,促进学科交叉与学科融合方面扮演着重要角色,是目前最活跃、发展最迅速、最有影响力的科学技术领域之一。航空航天科技的发展在一定程度上反映出一个国家的科学技术水平。

　　航空航天科技英语是科技英语的重要组成部分。以下分别阐述航空航天科技英语的文体特征、语言特点和翻译原则,其中涉及的词汇和句法特点普遍存在于航空航天文本中。科技英语翻译的基本策略和原则也适用于航空航天科技英语的翻译。

第1节　航空航天科技英语的文体特征

　　航空航天科技英语是科技英语的一个分支,兼具一般科技英语的文体特征和航空航天领域的专业特性。科技英语(English for Science and Technology,EST)是一种重要的英语语体,是专门用途英语(English for Special Purpose,ESP)中的一支,近几十年受到了国内外学者的广泛关注。科技英语可以泛指一切用于科学和技术交流的书面语和口语,内容涵盖自然科学和社会科学的诸多领域,涉及众多文体。

　　总体而言,科技英语文体讲求逻辑条理清晰和叙事准确严谨,在语言、词汇和语法方面,都有自己的鲜明特点。按照其适用领域,科技英语可以分为专用科技英语和普通科技英语。前者的题材和范围包括基础科学理论、技术性法律条文,涉及科学实验、科学技术研究、工程项目、生产制造等领域;后者一般用于传播科技知识、描写生产过程、说明产品使用方法等。区分专用科技文体和普通科技文体的一个重要指标是其语旨,即参加者的身份与他们之间的关系。专用科技文体的目标读者是专家(expert to expert writing),而普通科技文体是内行写给外行(scientist/journalist to layperson writing)的。前者的文体特征是表述客观、逻辑严谨、行文规范、用词正式、句式严谨;后者则用词生动、句法简易、文风活泼、多用修辞格。

一、专用科技文体特征

　　专用科技文体特征可以概括为语义客观、用词正式、结构严密、程式化。具体有以下几点：① 多用专业术语，包括缩略语；② 名词词组和名词化结构普遍；③ 多用被动语态；④ 句子长而复杂，句子成分多；⑤ 多使用后置修饰语和非限定性动词。

以下例文是选自 2013 年国际行星探测器研讨会的一篇论文摘要，作者是美国国家航空航天局埃姆斯研究中心的两名研究人员。

Mars Sample Return Landed with Red Dragon

A Mars Sample Return (MSR) mission is the highest priority science mission for the next decade as recommended by the recent Decadal Survey of Planetary Science. However, an affordable program to carry this out has not been defined. This paper describes a study that examined use of emerging commercial capabilities to land the sample return elements, with the goal of reducing mission cost. A team at NASA Ames examined the feasibility of the following scenario for MSR: A Falcon Heavy launcher injects a SpaceX Dragon crew capsule and trunk onto a Trans Mars Injection trajectory. The capsule is modified to carry all the hardware needed to return samples collected on Mars including a Mars Ascent Vehicle (MAV), an Earth Return Vehicle (ERV) and Sample Collection and Storage hardware. The Dragon descends to land on the surface of Mars using SuperSonic Retro Propulsion (SSRP) as described by Braun and Manning [IEEEAC paper 0076, 2005]. Sample are acquired and delivered to the MAV by a prelanded asset, possibly the proposed 2020 rover. After samples are obtained and stored in the ERV, the MAV launches the sample containing ERV from the surface of Mars. We examined cases where the ERV is delivered to either low Mars orbit (LMO), C3= 0 (Mars escape), or an intermediate energy state. The ERV then provides the rest of the energy (delta V) required to perform trans Earth injection (TEI), cruise, and insertion into a Moon-trailing Earth Orbit (MTEO). A later mission, possibly a crewed Dragon launched by a Falcon Heavy (not part of the current study) retrieves the sample container, packages the sample, and performs a controlled Earth reentry to prevent Mars materials from accidentally contaminating Earth.

The key analysis methods used in the study employed a set of parametric mass estimating relationships (MERs) and standard aerospace analysis software codes modified for the MAV class of launch vehicle to determine the range of performance parameters that produced converged spacecraft designs capable of meeting mission requirements. Subsystems modeled in this study included structures, power system, propulsion system, nose fairing, thermal insulation, actuationdevices, and GN&C. Best practice application of loads and design margins for all resources were used.

本文是一篇科技论文摘要，属于专用英语科技文本。科技论文摘要包含论文正文的重要信息和基本要点,结构严谨、语言精练、表达连贯、自成篇章。这些语言特征在本文中都有所体现。从词汇层面,本文专业词汇密度大,如使用了 capsule(太空舱)、launcher(发射器)、Trans Mars Injection trajectory(反式火星注入轨道)、Mars Ascent Vehicle(火星样品提升器)、Earth Return Vehicle(回地车)、nose fairing(头部整流罩)、thermal insulation(热绝缘)等;文中还使用大量缩略语,如 MSR、NASA、MAV、ERV、SSRP、IEEEAC、LMO、TEI、MTEO、GN&C、MERs 等;文中还多次用到名词化结构:use of emerging commercial capabilities 和 the feasibility of the following scenario;使用了非限定性动词,如 as recommended by、as described by、using SuperSonic Retro Propulsion;大量句子是被动句结构,如 However, an affordable program to carry this out has not been defined. The capsule is modified to carry all the hardware needed to…, Sample are acquired and delivered to…, After samples are obtained and stored…;大量使用后置修饰语,如 used in the study、all the hardware needed to…、samples collected on Mars、modified for the MAV、modeled in this study、capable of meeting mission requirements 等,包括介词短语 with the goal of;从句法角度看,文中也不乏长句的身影:The key analysis methods used in the study employed a set of parametric mass estimating relationships (MERs) and standard aerospace analysis software codes modified for the MAV class of launch vehicle to determine the range of performance parameters that produced converged spacecraft designs capable of meeting mission requirements。从文体层面看,段落通过诸如 However,then,after 等连接词实现较好的衔接和连贯,文章整体逻辑性强,叙述客观准确。

除此之外,作为科技论文摘要,本文采用了典型语句,例如破题方面,用主动句介绍作者或论文观点与内容:This paper describes a study that…,关于结论和建议也主要使用主动句。文中时态多用一般过去时,描述和总结已经完成的各项工作。

二、普通科技文体特征

普通科技文体主要包括科普文章和技术文本,其文体风格与专用科技文本有一定的区别。

以科普文章的文体为例。科普文体是科技文体的一个分支,目的是普及科学技术知识,科学性和文学性兼顾,知识性、通俗性和趣味性并重,语言上讲究通俗易懂、深入浅出,具体体现在用词生动、句法简易、文风活泼,多用普通词汇和修辞格。以下文为例。

The Beginning of Astronomy

Early men were also greatly interested in the stars they saw twinkling in the sky at night. What they did not realize was that the stars were also present during the daytime, but their light was not visible because of the brilliance of the sun.

How did they discover this?

There is one event which does enable us to see the stars in the daytime. It is the total eclipse of the sun, when the moon passes in front of the sun and hides it from view. When this happens, the sky becomes dark enough for the stars to be seen. Total eclipses of this kind do not occur very often and are not likely to be see from the same area more than once every fifty four years. One can imagine what an awe-inspiring sight it must have been for an early man, which would remember the event for the rest of his life.

So man watched the movements of the sun、the moon and the stars and wondered about them. It was practising the oldest branch of science—astronomy.

本文是一段关于天文学起源的叙述。文章从太阳、月亮和星星说起,对日全食景象进行了特别描绘。整篇文章偏口语化、通俗易懂,用语生动形象。文中使用了一些修辞手法(如用 twinkling 形容星光闪烁)和具有明显文学色彩的词(如 awe-inspiring, wonder about)等,增加了文学性,易于被读者接受。

技术文本指向特定读者传递技术信息的一种文本写作形式,其根本任务在于准确凝练地传递技术信息,其内容涵盖一切传递技术信息的文本类型和文本格式,包括技术报告、操作手册、商品目录、各种说明书、图表、广告宣传、招商材料、企业介绍、产品说明、信件、备忘录、建议书、商务策划书、可行性研究报告、培训方案、科技文献等。

以下例文摘自教材《活塞式航空动力装置》中的部分章节,主要讲述发动机操控中的注意事项。

Engine Handling

At all times, follow recommended procedures round in the manufacture's handbook. This will ensure correct operation of the engine, thereby avoiding spark plug fouling, avoiding over-stressing the engine components achieving best fuel economy and so on. Know the manufacturer's engine limitations and do not exceed them for reasons of safety.

When the engine is operating, you should monitor the oil temperature gauge (and the cylinder-head temperature gauge if installed) to help guard against high temperatures which are damaging go the engine. An abnormally high engine oil temperature could indicate insufficient oil in the engine-engines are especially dependent upon the circulation of the lubricating oil for their internal cooling.

High engine temperatures, either in the air or on the ground, will cause:

- loss of power;
- excessive oil consumption; and
- possible permanent internal engine damage.

In flight, you could consider cooling the engine by opening the air-cooling cowl flaps (if installed), richening the mixture, reducing power, or lowering the nose and increasing airspeed.

Avoid running the engine on the ground for prolonged periods if possible but, if unavoidable, face the aircraft into wind for better cooling and, if they are installed, open the cowl flaps. If the limiting red line temperatures are approached during ground operations, consider taxiing clear of the runway and shutting the engine down to allow cooling.

本文属于科技文体中指示类文体。这类文体可用于表达要求、建议、请求、命令、禁止等。在英语指示类文体中往往使用祈使句、情态动词以及否定词等,其结构严谨、叙事客观、用词正式。

本文使用了专业术语,如 spark plug fouling、oil temperature gauge、cowl flap、mixture、nose、taxiing。此外,本文主要描述发动机操作规程,其内容强调的是行为而非动作的执行者,所以文中采用较多的祈使句,省掉了主语或逻辑主语,简明扼要的祈使句结构可弱化感情色彩,体现客观性,增强文章的逻辑性,如 At all times, follow recommended procedures…;Know the manufacturer's engine limitations and do not exceed them for reasons of safety;When the engine is operating, you should…。文中还出现结构比较复杂的长句,有名词化结构以及非限制性动词等,如 This will ensure correct **operation of the engine**, thereby **avoiding** spark plug fouling, **avoiding** over-stressing the engine components **achieving** best fuel economy and so on. 和 An abnormally high engine oil temperature could indicate insufficient oil in the engine-engines are especially dependent upon **the circulation of** the lubricating oil for their internal cooling。此外,文中还使用了不少条件句,如 if possible …、if unavoidable…、If the limiting red-line temperatures are approached。以上都是典型的科技文体语言特征。

三、科技英语文体的新趋势

随着时代进步、科技发展以及文化多元化,科技英语的语言风格也在发生变化。如今,更加自然、简洁和朴实的文风备受科技界推崇。过去科技论文要避免过多使用第一人称句,因为会给读者造成主观臆断的印象,不符合强调客观准确的科技论文的写作原则,还会让读者觉得论文作者不够谦虚。如今,主流的观点认为被动语态比主动语态冗长、效率低,增加了读者的理解难度,采用第一人称句和主动语态可使句子更为简洁,不会削弱科技论文的客观性,反而使得表达更为清晰简洁,直接有力。

在英美等国家,越来越多的科技工作者在写作时选用第一人称和主动语态。许多世界权威的学术期刊(如英国的 Nature 和美国的 Science)都允许使用 I 和 we。Nature 在《作者投稿指南》中明确规定,论文摘要必须包含由"Here we show"或同类短语引导的概括主要结论的陈述语句。而使用第一人称 we 和像 conclude,suggest, argue,show 这样的主动动词搭配在科学界被视为是正常的、得体的。

在科技学术论文中使用第一人称和主动语态,可以强调作者的新成果和新结论,突出作者的学术观点,更好地与他人或前人的结论区别开。此外,科技论文摘要可以多采用以主动语态为主、被动语态句子为辅的语态形式,使主题更明确、句式更凝练、叙述更直接清晰。

请看以下例文:

Sub-scale flight test model design:Developments, challenges and opportunities

Growing interest in unconventional aircraft designs coupled with miniaturization of electronics and advancements in manufacturing techniques have revived the interest in the use of Sub-scale Flight Testing (SFT) to study the flight behaviour of full-scale aircraft in the early stages of design process by means of free-flying sub-scale models. SFT is particularly useful in the study of unconventional aircraft configurations as their behaviour cannot be reliably predicted based on legacy aircraft designs. In this paper, we survey the evolution of various design approaches (from 1848 to 2021) used to ensure similitude between a sub-scale model and its full-scale counterpart, which is an essential requirement to effectively perform SFT. Next, we present an exhaustive list of existing sub-scale models used in SFT and analyse the key trends in their design approaches, test-objectives, and applications. From this review, we conclude that the state-of-the-art sub-scale model design methods available in literature have not been used extensively in practice. Furthermore, we argue that one sub-scale model is not sufficient to predict the complete flight behaviour of a full-scale aircraft, but a catalog of tailored sub-scale models **is needed** to predict full-scale behaviour. An introduction to the development of such a catalog **is presented** in this paper, but the development of a formal methodology remains an open challenge. Establishing an approach to develop and use a SFT catalog of models to predict full-scale aircraft behaviour will help engineers enhance confidence on their designs and make SFT a viable and attractive testing method in the early stages of design.

本文出自国际航空航天领域顶级学术期刊 *Progress in Aerospace Science*,发表于 2021 年 12 月,是一篇科技论文摘要。全文由八个句子构成,其中只有两处使用了被动语态。文中第一人称主语频繁出现(we survey,we present,we conclude,we argue),多达四次。

国内有学者指出,在英美等国家,科技英语采用第一人称和主动语态已经在一定程度上成为主流,并对全球科技文献写作产生影响。中国在科技英语写作和翻译教学方面亟须了解和顺应这种变化,及时改革教学内容。广大科技工作者和专业翻译人员也有必要了解这一变化。

课后练习

1. 阅读以下科技论文摘要,注意其中的文体特征。

In-orbit demonstration of an iodine electric propulsion system

Propulsion is a critical subsystem of many spacecraft. For efficient propellant usage, electric propulsion systems based on the electrostatic acceleration of ions formed during electron impact ionization of a gas are particularly attractive. At present, xenon is used almost exclusively as an ionizable propellant for space propulsion. However, xenon is rare, it must be stored under high pressure and commercial production is expensive. Here we demonstrate a propulsion system that uses iodine propellant and we present in-orbit results of this new technology. Diatomic iodine is stored as a solid and sublimated at low temperatures. A plasma is then produced with a radio-frequency inductive antenna, and we show that the ionization efficiency is enhanced compared with xenon. Both atomic and molecular iodine ions are accelerated by high-voltage grids to generate thrust, and a highly collimated beam can be produced with substantial iodine dissociation. The propulsion system has been successfully operated in space onboard a small satellite with manoeuvres confirmed using satellite tracking data. We anticipate that these results will accelerate the adoption of alternative propellants within the space industry and demonstrate the potential of iodine for a wide range of space missions. For example, iodine enables substantial system miniaturization and simplification, which provides small satellites and satellite constellations with new capabilities for deployment, collision avoidance, end-of-life disposal and space exploration.

2. 阅读以下科普文章,注意其中的语言特点。

In the 400 years since humans began studying the cosmos with telescopes, Saturn has been transformed from a faint and blurry spot in the sky to a colossal planet in sharp focus. Yet, it seems that the clearer we see Saturn, the more enigmas we find.

Saturn is 763 times the volume of Earth with a primarily hydrogen and helium atmosphere that's like a wild, churning cauldron. "Most of the weather on Saturn is erratic," said Andrew Ingersoll, a member of the Cassini imaging team and an atmospheric physicist at Caltech. "A year will go by with no thunderstorms at all, but every 20 to 30 years you get a giant storm flashing with lightning every tenth of a second." Cassini watched that mega-storm erupt in 2010, sending atmospheric disturbances all the way around the planet until the storm began running into its own wake, appearing to eat its own tail before it faded and disappeared.

Not all of Saturn's characteristics come and go, though. "Some features are surprisingly longlived," Ingersoll said. "Cassini observed Saturn's north polar hexagon, discovered in 1981 by Voyager, and it's big enough to contain two Earths. And each pole sports a hurricane-like 'eye' that has been around for a decade or more. These are special properties, unique to Saturn, and they are showing us how extreme the weather can be."

Further, scientists don't know with precision how long it takes for Saturn to complete a single rotation on its axis (i. e., the length of a Saturn day), or why the planet's magnetic poles are aligned with the axis of rotation—the only planet known to have such an alignment. The size of Saturn's core and the planet's internal structure are also practically unknown.

第 2 节　航空航天科技英语的语言特点

科技英语文体结构严谨、条理清晰、行文规范、描述客观、语气正式,在词汇和句法方面有比较鲜明的特征,其主要特点是专业术语、缩略语、专有名词、数字、符号和公式使用频繁,用词简洁精准。句子多采用名词化结构、被动句式、非谓语动词结构以及结构比较复杂的长句等。

一、专业术语的使用

术语(terminology)是表达或限定科学概念(concept)的约定性语言符号,是交流思想和认识的工具。科技专业术语指在自然科学、社会科学和工程技术等学科领域用来表达特定概念的词汇和词汇组合。专业术语反映科学认识的成果,是科技信息的载体,保证了科技文章中概念的规范性和准确性,具有高度的概括性。科技文章正是依靠这些专业术语来表达其科技内容的,并使科技信息的传达和交换准确化、精确化。

专业术语一般具有单义性、稳定性、系统性和能产性等特点。单义性指的是一个科学术语只能表达一个概念,反过来一个概念也只能用一个科学术语表达,不容许有歧义。与一般字词有多种含义不同,专业术语在某个特定专业范围内是单义的,即使有些术语应用于多个专业领域也仍然具备这个特征。如:airfoil(翼面)、turbine(涡轮)、spacecraft(航天器)、standard star(标准星)等。稳定性指术语一旦规范、定名,一般会长久使用,不轻易更改。系统性指在特定领域,每个术语的地位只有在其整个概念体系中才能显现,各个术语必须处于一个明确的层次结构中,共同构成一个系统。能产性指确定后的术语可以通过构词法等派生出新的术语,如 aero-(航空的、飞行的、飞机的)可以构成许多与航空相关的词汇:aeroplane(飞机)、aerospace(空间)、aerodynamics(动力学)、aerodyne(重于空气的航空器)等。

科技词汇包括纯科技词汇(专业科技词汇)、半科技词汇和通用科技词汇。纯科技词汇指专门指称某一专业领域的词汇。这类词汇的语义单一、固定、规范、严谨,国际通用,对语境的依赖度低,与其他词汇相比,出现频率较低。如:航空航天领域中的 hatch(舱口)、astronaut(航天员)、access flap(接口盖)、low Earth orbit(近地轨道)、lunar rover(月球车)、payload capability(有效载荷能力)、experimental spacecraft(试验太空船)等。

半科技词汇(次技术词或准术语)指科技英语中使用的普通词汇。这些词汇既有普通词义又有科技词义,其科技词义是由普通词义扩展而来的,而且在不同的专业领域具有不同的专业词义。科技英语中的这类词汇词义繁多、用法灵活、搭配形式多样、使用范围广,与其他词汇相比,出现频率非常高。如:nose(机首)、ceiling(升限)、ferry range (转场航程)、taxi(滑行)、apron(飞机库前跑道、停机坪)、stroke(冲程)、pencil rocket (高空气象观测用的小型火箭)等。

通用科技词汇指不同专业都要经常使用的通用词汇,这类词词义比较单一,使用范围比专业词汇要广一些,在科技文本中出现频率较高,如:energy(能量)、frequency(频率)、density(密度)等。值得注意的是,通用专业词汇在不同专业中的语义日趋多样化。如 carrier 一词,在航天领域表示"载波",在航空领域表示"运输机",在军事领域表示"航空母舰";再如 integrity 一词,在数学、计算机领域中被译为"完整性",在航空科学和导航专业中被译为"完好性(又称完善性)",前者强调数量,后者数量和质量兼顾,或偏重数据的品质。

航空航天科技集大成于一身,涉及的科学技术领域极其广泛,其专业术语数量多、专业性极强。航空航天科技在迅猛发展的同时涌现出大量的科技新词。如英语新闻中用来专指中国航天员的 taikonaut,借用希腊和拉丁词素的 aerospace(航空与航天空间)、缩略语 VTOL(vertical take-off and landing 垂直起落飞机)、拼缀词 lunarnaut(lunar＋astronaut,月球宇航员)、lunarcast(lunar＋telecast 登月电视广播)、复合词 comsat (communication satellite,通信卫星)、moonquake(月震)、moonwalk(月球漫步)、probe(用于科学考察的宇宙飞船)和外来语借入词 sputnik((俄语)人造卫星)、kamikaze((日语)遥控飞行器)等。

此外,航空航天科技术语的相关词汇表意精细,在翻译时一定要注意区分,准确辨析词义。例如 module、unit 和 assembly 都有"组件"的意思,但在航空词汇中,这三个词对应的是不同的概念:module 常指"可视面板组件",如飞机驾驶舱内的仪表;unit 主要指电气和电子设备上的组件,如 generator control unit(GCU)(发电机控制组件);assembly 常用在机械专业领域,指由不同零件构成的一个整体,如飞机机轮组件。再例如 cabin 一词,《航空科学技术名词》中译为"座舱",《交通大词典》则译为

"机舱"。其实这两个词是并列词,前者对应小型机,后者对应大型机。

科技语言区别于其他语体的最显著标志就是科技专业术语。科技专业术语的翻译是科技翻译的难题之一,特别是在航空航天这样的前沿科技领域,新技术、新发现会催生更多的新术语,给翻译带来新的挑战。

二、名词化结构的使用

科技英语中的名词化倾向显著,表现为名词化结构使用广泛。名词化结构指将动词或形容词转换为名词或名词词组,使之在语法上具备名词的功能并能在句子中充当名词使用,语法功能往往与从句相当。动词名词化的结果是科技英语中行为名词大大增加。这些词通常由动词加-sion、-xion、-ment、-ance、-ence、-age、-cy、-ty、-ery等后缀构成。这类名词除表示行为动作外,还可表示状态、手段和结果等。名词化结构在科技英语中主要有以下几种形式:从动词或形容词转化的抽象名词、具有显著动作意义的实施者名词、搭配介词的名词化结构等。名词化结构通常以短语的形式出现在句子中,具有言简意赅、组织严密、信息容量大、各部分之间的逻辑关系明确等特点,适宜表达精细复杂的思维逻辑,高度符合航空航天科技文本要求语言规范、语气正式、陈述客观、条理清晰、逻辑性强的文体特征。

以下句子中的划线部分为名词化结构:

1. The attainment of hypersonic speeds is now quite common.（动词转化的名词化结构）

2. Lowe's experiments with balloons led him to believe in the existence of an upper stream of air that moved in an easterly direction，no matter what direction the lower currents flowed.（动词转化的名词化结构）

3. The miniaturization of electronics and COTS components in the recent years has opened up avenues for sub-scale model designs whose size can now vary over a range from 50 cm span（with micro-measurement devices）to over 4 m（with powerful yet small jet or electric engines）.（动词转化的抽象名词）

4. Flight testing of a prototype（the physical device or system for which the predictions are to be made i. e. , a full-scale aircraft in aerospace applications）is the best way to ascertain the flight behaviour of a radical aircraft design.（动词转化的名词化结构）

5. Well into the 1970s，one of the most popular and influential geological textbooks，The Earth by the venerable Harold Jeffreys，strenuously insisted that plate tectonics was a physical impossibility，just as it had in the first edition way back in 1924.（形容词转化的抽象名词）

三、被动语态的使用

英国著名语言学家夸克（R·Quirk）在《当代英语语法》（*A Grammar of Contemporary English*）中提到科技英语的两个最显著的特点是广泛使用名词化结构和大量使用被动语态。科技英语中被动语态的使用率比主动语态高出十多倍。

航空航天科技英语用以描述和记录与航空航天有关的科技知识和科技活动，强调客观准确，侧重叙事推理，因此较多采用被动语态。此外，被动语态的使用使得文章句式结构更加灵活，有利于整体衔接和连贯。以下是科技英语中被动语态使用的例子。

1. Human space missions <u>are now conducted</u> for 6-month durations, and a one-year mission <u>is planned</u> for 2015.

2. In the past, as an alternative to unaffordable full-scale aircraft, sub-scale models <u>have been used</u> for flight testing.

3. Boeing <u>is contracted</u> to build two Core Stages and is also responsible for integrating and testing the Core Stage with the four RS-25 engines and government-provided flight control software.

4. A rubber ball containing a radio transmitter <u>could be dropped</u> from a satellite so that it would fall towards the surface of the planet. The radio would signal the rate at which the ball <u>was slowed</u> down and scientist would be able to calculate how dense the atmosphere is. It may even be possible to drop a capsule containing scientific instrument on to the planet's surface.

四、非限定性动词的使用

科技英语中大量使用分词、动名词和动词不定式等非限定性动词（nonfinite verb）。这些语言形式不受主语的人称、数和时态的限制，可以准确地反映句子各部分的内在关系。非限制性动词的使用可以使科技英语文本行文简练、句子结构紧凑、逻辑性更强。以下划线部分是非限制性动词。

1. Aeroplanes can be powered by a variety of engines, the two fundamental types <u>being</u> piston engines and gas turbines (jets).（分词结构 being piston engines and gas turbines (jets)具体说明 the two fundamental types 的内容）

2. Today the electronic computer is widely used in solving mathematical problems <u>having to do with</u> weather forecasting and putting satellites into orbit.（分词短语 having to do with 修饰先行词 mathematical problems）

3. Ideally, a way would be found to allow supersonic speeds at low altitudes, <u>further enhancing the aircraft's survivability.</u>（分词短语 further enhancing the aircraft's survivability 作目的状语）

4. In 2001, the use of ion propulsion was evaluated by JPL and it was determined that the use of state-of-the-art electric propulsion, the NSTAR thruster, could be used to enable a single launch architecture <u>using electric propulsion for the transfer to Mars, orbit insertion and the return.</u> （分词短语 using electric propulsion for the transfer to Mars, orbit insertion and the return 作方式状语）

五、后置修饰语的使用

后置修饰语（postmodifier）是科技英语中极为常见的一种语言现象。形容词、形容词短语、介词短语、关系分句、同位语分句、不定式短语、现在分词短语、过去分词短语均可以作后置修饰语。以下划线部分是后置修饰语：

1. In particular, the development of a detailed component performance and weight model, <u>applicable to</u> the analysis of semi-closed cycles and open cycles, is briefly presented. （形容词短语作后置修饰语）

2. All commercially funded product development projects <u>of high complexity</u> share a common dilemma. （介词短语作后置修饰语）

3. Its capacity <u>to penetrate air defenses and threaten effective retaliation</u> provides a strong, effective deterrent and combat force well into the twenty-first century. （不定式短语作后置修饰语）

4. The location of the wing along the longitudinal axis of the fuselage <u>relative to the center of gravity of the complete aircraft</u> is determined by stability and control requirement in pitch. （形容词短语作后置修饰语，修饰 the location of the wing）

六、复杂长句的使用

科技英语常使用结构比较复杂的长句，目的是严谨准确地表述复杂的概念，并使句子结构紧凑，逻辑严密。这些句子修饰成分多，环环相扣，理解时需谨慎对待。汉语和英语在结构以及语义表达上存在比较大的差异，翻译这些结构复杂的长句时需要分步骤，找出句子的中心内容，厘清各部分之间关系，按照汉语的特点和表达方式，采用适当的翻译策略进行转换。以下是一些结构复杂的长句：

1. Severe budget cuts to the human research and development budget, coupled with a growing awareness that organizations outside of NASA were advancing human health and performance innovations at an increasingly rapid pace, drove the NASA Space Life Sciences Directorate to adopt a strategic plan in 2007 grounded on a new business model anchored in collaboration.

2. With Astronaut John Young piloting the 75 ton Columbia with a stick and

pedal manually like a giant glider, the shuttle came to a halt on a dry lake bed on the desert air base after a 54-hour space flight in which it circled the Earth 36 times.

3. When NASA first built the space shuttle and with it the first chance to carry both man and equipment inexpensively into space, it opened the doors to the eventual building and operating of commercial shuttles by private corporations that, possibly within the next generation, may be able to fly us to a space as routinely as we are flown to Europe.

4. The construction of such a satellite is now believed to be quite realizable, its realization being supported with all the achievements of contemporary science, which have brought into being not only materials capable of withstanding severe stresses involved and high temperatures developed, but new processes as well.

七、有限的时态种类

科技英语中出现的时态有限,常用的有一般现在时、一般过去时、一般将来时和现在完成时。

1. 一般现在时用来叙述过程、客观事实、科学定理、通常或习惯性的行为以及一般性结论。如：

The earth is one of the nine planets. They all go around the sun the same way. Each of them also turns on its own axis. The earth makes a complete journey round the run in one year. It turns once on its axis in 24 hours. The sun rises and sets because the earth is turning.

2. 一般过去时用来介绍过去的科研活动,如实验和发明创造等。如：

The Spirit of Kansas, AV-12 (89-0127), taxis on the Whiteman ramp the summer of 2005. Its first flight was on December 5, 1994, and it was delivered to the 509th Bomb Wing on February 17, 1995. It was named on May 13, 1995.

3. 一般将来时用来叙述计划要做的工作、预期获得的结果和规律性的倾向等。如：

There will be atomic jetplane, we will be able to fly right up through the atmosphere, through the stratosphere, and away out into space, to the moon, and Venus and the other planets.

4. 现在完成时表示到目前为止发生的行为、对现在产生影响的行为等。如：

Since 2009, NASA has utilized collaboration and open innovation processes to seek solutions to some of the more difficult and pressing space medical care issues. These novel approaches have resulted in significant advances to achieving NASA's human health and performance goals, as well as benefiting life on Earth.

课后练习

一、将下列科技术语翻译成汉语。

1. seaplane
2. avionics
3. turbofan
4. trainer aircraft
5. angel view
6. ATC
7. tubeless tyre
8. gear box
9. landing gear
10. bypass valve
11. avionics specialist
12. throttle lever
13. field support

二、将下列句子翻译成汉语,注意其中的语言特点。

1. It requires the identification of fuel tank systems design facial appearance that may lead to development of ignition sources in it.

2. Instruments capable of recording, with reasonable accuracy, the maximum pressure and muzzle velocities have been available for a number of years.

3. Part of the compactness benefit also comes from the significantly higher overall cycle pressure ratio achievable with a semi-closed cycle configuration, leading to a significant reduction of the low pressure components and the recuperator weight.

4. For many years, NASA and the science community have been asking for a Mars Sample Return（MSR）mission. There have been numerous studies to evaluate MSR mission architectures, technology needs and development plans, and top-level requirements.

5. After Galileo's work the feeling grew that there were universal laws governing the motion of bodies and that these laws might apply to motion in the heavens as well as on earth.

6. The electric propulsion architecture is attractive because it offers a single launch solution on a medium class EELV with very large launch windows, a few

month.

7. If the aerospace industry needs to shift towards unconventional configurations, the ability to accurately evaluate the aircraft flight behaviour in the early stages of design cycle is essential not only to prevent costly last-minute rework but to assess their viability too.

8. In 2010, NASA established the NASA Human Health and Performance Center (NHHPC) as a forum to exchange best practices in innovative approaches, and to facilitate partnership development across sectors for solving medical, environmental, and human performance problems in space and on Earth.

9. The Michelson-Morley experiment was carried out in 1887 to determine the motion of the earth through the ether, a presumed absolutely motionless, invisible, tasteless, and odorless substance that permeates everywhere of the universe so that light could have something to propagate through.

10. Given the diverse nature of the space flight population and increasing length of missions, ever more sophisticated medical care and environmental health systems and an infusion of novel approaches and technologies to improve health care in flight are required.

第3节 航空航天英汉翻译概述

科学技术是第一生产力。作为服务全球科技文化交流和社会发展进步的科技翻译，无疑是推动科技进步的重要因素之一。航空航天技术是人类探索宇宙空间的新兴领域，是一个国家综合实力的反映，在推动国家经济发展、提升我国国际竞争力方面发挥着重要作用。从中国传统翻译理论的"目的性"角度看，科技翻译担负着介绍国外先进科学技术和把握国外科学技术动向的重任，已经成为科研活动不可或缺的组成部分。

一、科技翻译的特点

科技翻译的对象主要是科技领域(如航空、航天等)的各类文本，既包括对具有技术、工艺、科学性质的纯科技文本的翻译，也包括对兼具科学性和文学性的科普文本和科幻作品的翻译，因此科技翻译的读者对象不局限于科技人员，还包括非专业人士和普通读者。科技翻译属于应用文体翻译。应用文体翻译主要是指与文学翻译相对的实用文体翻译，如商务翻译、政治翻译、法律翻译和影视翻译等。

有学者将科技翻译的特点总结为客观性、规范性和科学性。科技翻译的客观性是由科技文本的特点决定的。科技文本涉及科学事实、科学理论、科技成果报道、科学活动和信息交流。在进行科技翻译时，译者必须尊重科学事实，对翻译的内容永远保持客观的态度。这具体表现在译者不能任意篡改原文，或者回避原文中自己不懂的地方，更不能在译文中随意添加主观臆测或评论。科技翻译的规范性指翻译中涉及的专业术语都是各学科领域的专门用语，已经严格规范化、为学界普遍接受。科技翻译的科学性有多重解读。首先，科学性指科学是科技翻译存在的前提，没有科学便没有科技翻译。同时，在翻译时尊重并反映科技文体的特点也是科技翻译科学性的一种表现形式。其次，科技翻译人员的翻译过程同样也体现了一定的科学性：从对原文的内容、形式、风格和语言的分析到用译入语对原文的创造性再现过程，无一不涉及人类思维的逻辑规律，而科学的本质正体现在其中。

二、科技翻译的标准

不同类型的文本应遵循不同的翻译标准，而采用适合的翻译原则和方法是当代翻译理论的核心思想。科技文体按照专业程度和适用目标读者群，可以分为专用科技文体和普通科技文体。两者的翻译标准既有共通之处，也存在一些差异。

（一）专用科技文本的翻译标准

专用科技文本一般语言规范、语气正式、陈述客观、逻辑性强、信息量大、高度专业化。科技翻译在译文中要尽可能体现这些特点，译文要力求内容准确、结构严密、概念清楚、语句达意。

"信"和"达"，即"忠实"与"通顺"，是公认的翻译标准，同样适用于科技翻译。需要强调的是科技文体专业性强、表达严谨客观，因此"信"在科技翻译中尤为重要。科技翻译中的"信"指译者必须准确且完整地理解原文的全部内容，不能任意增删、曲解、篡改或遗漏。科技文体的交际目的是将科学技术的研究成果和记录应用于指导实践，翻译转换过程中对原文信息的任何遗漏、增添或偏离都可能带来严重的后果。康志洪在《科技翻译》一书中指出，科技翻译从总体技术难度上并不比其他类型的翻译更难，但科技翻译的质量在实际应用受到检验的严苛程度是其他类型翻译无法超越的。因此，科技翻译的译文必须准确，不能模棱两可，以免产生歧义。准确性和客观性是科技翻译的核心。

科技翻译中的"达"要求译者必须考虑科技文体的特点，遵循科技翻译的目的，译文既要用地道的译入语表述，要通顺易懂，避免文理不通、结构混乱或逻辑不清，同时还必须符合学科要求和专业表达习惯，采用规范的专业语言。这里的"规范"和"专业"指译文的专业术语表述要符合科技语言和术语的规范，尽可能用译入语中已有的约定俗成的定义、术语和概念。方梦之认为，科技译文就要翻译得像科技文章，不但内容是科技的，而且遣词造句、文风墨气也是科技的。以下举例说明。

Proof of compliance

（a）Each requirement of this subpart must be met at each appropriate combination of weight and center of gravity within the range of loading conditions for which certification is requested. This must be shown：

① By tests upon an airplane of the type for which certification is requested，or by calculations based on，and equal in accuracy to，the results of testing；

② By systematic investigation of each probable combination of weight and center of gravity，if compliance cannot be reasonably inferred from combinations investigated.

(b) Reserved

(c) The controllability, stability, trim, and stalling characteristics of the airplane must be shown for each altitude up to the maximum expected in operation.

(d) Parameters critical for the test being conducted，such as weight, loading (center of gravity and inertia)，airspeed，power，and wind，must be maintained within acceptable tolerances of the critical values during flight testing.

译文　证明符合性的若干规定

(a) 本部分的每项要求，在申请审定的载重状态范围内，对重量和重心的每种相应组合，均必须得到满足。证实时必须按下列规定：

① 用申请合格审定的该型号飞机进行实验，或根据实验结果进行与实验同样准确的计算；

② 如果由所检查的各种组合不能合理地推断其符合性，则应对重量和重心的每种预期的组合进行系统的检查。

(b) 保留

(c) 必须显示飞机的可控性、稳定性、配平和失速特性，直至达到预期的最大运行高度。

(d) 对进行测试的关键参数，例如重量、负载(重心和惯性)、空速、功率和风，在飞行测试期间必须保持在临界值的可接受公差范围内。

以上译例原文选自美国联邦航空管理局颁布的 PART 25-AIRWORTHNESS STANDARDS：TRANSPORT CATEGORY AIRPLANES (运输类飞机适航标准)，属于行业法规，文本的功能以表达为主，文体庄重、语言正式、句子语义明确、结构紧凑。翻译时要注意这些语言特点，做到译文严谨、准确和等效。

(二) 普通科技文本的翻译标准

普通科技文本主要包括技术文本和科普文章。上文提到的"信"和"达"，也适用于普通科技文本。但是，科普文本的翻译标准有一些特殊之处。科普作品的翻译标准主要由科普著作的性质和特点与科普著作的目标读者群决定。按照《中华人民共和国科学技术普及法》的定义，科普作品是科技文体的一种变体，是文学和科学相结合的写作体裁，其目的是"普及科学技术知识、倡导科学方法、传播科学思想、弘扬科学精神"。郭建中将科普文本的特点概括为科学性、文学性、通俗性、趣味性。科普著作的目标群体广泛,任何对科学知识感兴趣的人都可能是科普作品的读者,他们中既有成年人也有儿童。这些特点决定了科普翻译必定是科技文本翻译与文学文本翻译的结合体。科普翻译的译文,首先要保证正确性,要忠于原意,正确传达原文的知识脉络,体现原文内容的科学严谨性;其次,要忠实再现原文的文学性,文从字顺,表达要有文采。此外,科普翻译还要注意译文的通俗性和趣味性,行文要通俗易懂,富有情趣。科普翻译既不能为了美学目的损坏原文的科学性,也不能为了科学性而抹杀原文的艺术性和趣味性。只有两者兼顾才能为读者所接受和喜爱,才能最大限度地传播科学技术知识,真正达到科普的目的。

下面各例由于理解偏差或表达不当,在"信"和"达"方面存在一些问题。

1. For ranges of <u>up to</u> 500 or 600 miles, the turbo-prop can deliver its good fuel economy.

原译　航程<u>为</u>500英里或600英里(805千米或966千米)时,涡桨发动机飞机就能体现出其燃油经济性的优势。

改译　航程<u>长达</u>500英里或600英里(805千米或966千米)时,涡桨飞机才能体现出良好的燃油经济性。

分析　此句中的"up to"表示的是一个范围的上限,而非一个具体的(时间或距离)点。

2. Contrary to what is often thought the earth does not move round in the empty space.

原译　与通常的想法相反,在空无一物的空间里,地球是不旋转的。

改译　与通常的想法相反,地球并不是在空无一物的空间旋转的。

分析　错译的原因是译者忽略了自然常识。

3. You may have read stories in which the hero gets into his spaceship and "blast off" into outer reaches of space without a worry about fuel.

原译　你也许读过这样的故事吧。故事中的主人公乘坐飞船,发射升空,一点都不用担心燃料问题。

改译　你可能读过这样一些故事吧。这些故事描写主人公坐上飞船,"腾空而起",直冲九霄云外,无须为燃料担忧发愁。

分析　作为科普翻译作品,原译表达过于平淡,缺乏文采,改译运用了汉语四字格"腾空而起""九霄云外"和"担忧发愁",使译文形象生动,可读性大大提高。

4. The exploration of space, world finance and the development of new sources of power, such as atomic energy—these are examples of areas of scientific research which are so costly and complicated that no single country or organization, working by itself, can hope to tackle efficiently.

原译　对于空间和世界金融的探索,以及对于像原子能这样的新能源的发展——这些都是科学研究领域的例子,它们如此耗费昂贵和复杂,没有一个单独的国家或机构如果自己单干,能够希望有效地解决。

改译　空间探索、世界金融和新能源(如原子能)开发,都是耗资巨大而又十分复杂的科学研究领域的项目,任何一个国家或组织都不可能指望独自高效承担。

分析　句中的 world finance 和 the exploration of space 是并列关系,并非介词 of 的宾语,因此和 the exploration 并不构成限定关系。原译的表达信息割裂,读起来不够通顺,翻译腔比较浓。

5. Structurally, there is a requirement for good visibility from the flight deck in the nose section and a need for a kick-up of the bottom of the tail cone section to provide ground clearance during takeoff rotation.

原译　从结构上来说,要求机头部段的驾驶舱应有良好的能见度,而尾椎段的底部能够向上弯曲,这样在起飞旋转期间就可以保证离地净高。

改译　从结构上来说,要求机头部段的驾驶舱视野良好,而尾椎段的底部上翘,这样在起飞抬前轮期间就可以保证离地高度。

分析　原句使用了多个专业术语,如 visibility(视野良好)、kick-up(上翘)、ground clearance(离地高度)、rotation(抬前轮),这些专业术语都有约定俗成的汉语译法,而原译未能采用译入语的专业术语来表达。

三、科技翻译的译者素质

英国语言学家夸克(R·Quirk)曾指出,人们在试图讲授一种范围限定的语言(例如工程师用英语)时,往往没有看到这种做法的危险性,就好像是试图攀登陷在烂泥中的梯子一样,如果没有基础,即使设法爬到最高一级,也是徒劳无益的。在科技翻译中,夸克所说的"基础"就是指译者素质。翻译活动是一项高度复杂的综合性脑力活动,要求译者具有比较全面的综合素质,包括译者的知识面、双语修养、思维能力、表达能力等众多因素。科技翻译译者的基本素质主要包括以下三点:

(一) 扎实的双语基础

熟练掌握外国语和本族语是从事科技翻译工作的基本条件,是正确理解原文的思想内容,领会原文的风格和神韵,然后忠实流畅地转化为本族语的基础。俄国作家杜勃罗柳波夫曾说:"他(译者)要翻译的语言不言而喻应该是他精通的,精通词在意义上、在配置上或在添加一个不重要的语气词时所产生的一切细微的差别。不仅如此,译者还应极其熟练地掌握译文语言。他不仅要写得正确而优美,而且要写得轻松和流畅;在确切地表达原文的思想时,使读者看不出一点生硬的地方,能在翻译中运用语言的全部财富,不能使用不正确的措辞和不精准的词汇。"

（二）全面的知识结构

全面的知识结构主要由翻译和科技两个层面构成。虽然不是每一位译者都需要系统全面地学习翻译专业的所有理论知识，但必须具备基本的翻译常识，懂得翻译的基本原则和技巧。同时，作为科技翻译工作者，掌握一定的科技专业知识，具备基本的科学素养也是非常重要的。

（三）必要的信息技术

突飞猛进的信息技术在翻译实践中的运用也越来越广泛。翻译过程中的不少环节甚至可以直接由计算机来完成，比如专业词汇的记忆、术语的转化等。文本处理和分析软件以及计算机辅助翻译软件等信息技术搭建的翻译平台，高效地为科技翻译人员提供包括专业术语在内的多种参考译法，极大地改善了翻译环境，提高了翻译效率和质量。科技翻译人员应该顺应时代发展，掌握一定的信息技术，为更高效地完成高水准翻译工作打下坚实基础。

课后练习

将下列句子翻译成汉语，注意译文的准确和通顺。

1. A nearby object falling into a black hole is never heard from again.

2. Thus regardless of whether you are making a basic design or inventing your own there is only one way to tell if the kite is satisfactory: fly it.

3. He observed that the times at which the moons of Jupiter appeared to pass behind Jupiter were not evenly spaced, as one would expect if the moons went round Jupiter at a constant rate.

4. All in all he replaced the proud structure of nineteenth-century science, which had been so battered by the sensational discoveries at the turn of the century, by a new one whose foundations and framework were solid enough to withstand the fiercest storms which might come in the future.

5. A rubber ball containing a radio transmitter could be dropped from a satellite so that it would fall towards the surface of the planet. The radio would signal the rate at which the ball was slowed down and scientist would be able to calculate how dense the atmosphere is. It may even be possible to drop a capsule containing scientific instrument on to the planet's surface.

第 2 章　航空航天科技英语词汇翻译策略

词汇是构成语言的基本单位,数量众多,且其基本含义也在不断地发展变化。词汇翻译是航空航天科技英语翻译的基础。在翻译过程中,不仅要了解航空航天科技英语词汇的特点及其含义的发展变化,而且要掌握在特定语境中的翻译策略。

本章以航空航天科技英语词汇翻译为重点,详细阐述词义选择、词类转换、词义引申、词的增译和省译等翻译技巧,并简要介绍航空航天科技英语专业术语翻译的基本方法。

第 1 节　词义选择

航空航天科技英语中大量地使用科技词汇,其大致可以分为三类:专业术语词汇、半专业术语词汇和普通词汇。专业词汇通常反映客观事实,无感情色彩,有严格的语义界限,往往是单义词,常仅用于航空航天学科。如:satellite(卫星)、spacecraft(宇宙飞船)和 astronaut(航天员)等。半专业词汇指由普通词汇转化演变而来的一些除了本身含有的一般词义外,还在不同学科和专业中具有不同专业含义的多义词汇。如 power 一词,它的基本含义为"力量、权力",但在物理学专业可以释义为"功率",在天文学可以释义为"倍数",电学领域意为"电力"。在航空航天科技英语文章中,半专业术语词汇和普通词汇的使用率远远高于专业术语词汇,故本章主要讨论普通词汇和半专业术语词汇的翻译方法。随着航空航天技术的高速发展、新技术门类的出现、新概念的不断产生,新词汇大量涌入。在这种情况下,如何与时俱进、准确翻译该领域的词汇成为航空航天科技英语学习者的必备技能。

一、词义辨析

译者在航空航天英汉翻译中面临的首要难题是词义辨别。所谓词义就是词语在具体语境中的含义,是语篇中不可缺失的成分。正确的词义选择既要结合词典查阅,更要通晓语篇,结合上下文进行判断。在航空航天科技英语中,同一个词或词组在不同的上下文、不同的专业领域以及不同的语法结构中可能分属不同的词类、具有不同的词义,而汉语的词汇含义相对单一、固定。因此,译者在翻译时要根据语境、专业领域等因素,在不同的词义中选择出一个最恰当的词义。

（一）根据词类选择词义

在科技英语中，一词多类和一词多义现象比较普遍。一词多类指一个单词分属几个不同的词类；一词多义指同一个词在同一词类中具有几个不同的词义。在众多词义中，选择出一个最恰当的词义，是正确理解原文的基础。在科技翻译过程中，译者遇到一词多类和多义时，首先应根据句法结构判明这个词在原句中属于哪一词类，起什么作用，然后再根据词类选择一个恰当的词义。例如，base 一词分属于不同的词类时，其词义也会发生变化：

1. Plastics was at first based on coal and wood.

最初塑料是从煤和木材中提取的。（动词）

2. The machine rests on a wide base of steel.

机器安装在一个很大的钢制底座上。（名词）

3. Vodafone increased its customer base by 4.2 million.

沃达丰公司的客户群人数增长了 420 万。（名词）

4. All employees at the submarine base requires security clearance.

所有潜水艇基地的雇员必须得到安全部门的审查许可方可录取。（名词）

5. A transistor has three electrodes, the emitter, the base and the collector.

晶体管有三个电极，即发射极、基极和集电极。（名词）

6. The vigesimal system (counting in base 20) is thought to originate from human's fingers and toes.

据猜测，二十进制（以 20 为基数进行计算）源于人类的手指和脚趾。（名词）

7. Consequently this last base metal 50 Schilling issue commemorates the Schilling, which was the national currency of Austria for the greater part of the 20th century.

因此，此次发行的最后一枚 50 先令普通金属币是为了纪念"先令"，它在 20 世纪大部分时间是奥地利的法定货币。（形容词）

（二）根据上下文选择词义

上下文又称语境（context），即语言使用的环境。词语的意义由语境决定，脱离了语境，很难确定语义。语境在很大程度上决定了科技英语理解的准确性和译文是否通顺。科技英语词义比较灵活，需要借助具体语境提供的线索进行分析、推理和判断。以 develop 一词为例，在不同的上下文中，该词具有不同含义：

1. Noises may develop in a worn engine.

引擎磨损可能会产生各种噪声。

2. We do not develop software to make it object oriented，or to comply with any other paradigm.

我们开发软件并不是针对特定对象，也不是遵循其他的模式。

3. After the war，both the United States and the Soviet Union began to develop large rockets for exploring space.

战后，美国和苏联都开始研制大型火箭，进行太空探测。

4. We would also like to hear the opinions of our stakeholders and work together to develop our visa services further over the coming years.

我们也将听取各相关单位的意见，在未来几年共同努力，进一步提升我们的签证服务。

5. If a cataract blocks light from entering the eye and stimulating the retina，the area of the brain used for sight does not develop properly.

如果白内障阻碍了光线进入眼内刺激视网膜，那么大脑的视觉皮层就不会正常发育。

下面再以"flight"一词为例，说明如何根据上下文确定词义。

1. Airplanes retract their landing gear while in flight for the same reason.

出于同样原因，飞机在飞行时收回起落架。

2. These small booklets would inform passengers of flight schedules，security regulations，in-flight entertainment and food menus.

这些小册子向旅客提供了航班时刻、安全规定、机上娱乐和饭菜食谱等信息。

3. Modern flight simulation technology includes two major domains：flight simulation and in-flight simulation.

现代飞行模拟技术包括飞行模拟和飞行中模拟两个主要领域。

从以上例子可以看出，同一个词在不同的语境中往往具有不同的含义。在翻译时，译者必须要弄清这个词在不同上下文中的确切含义，再选择恰当的词义翻译。脱离了上下文，孤立地去理解词义，往往会导致译文错误。

(三) 根据专业特点选择词义

航空航天科技英语中，同一词的多种意义往往适用于不同的学科和专业。因此，译者在选择词义时，除了要掌握其基本含义，还应了解文本所涉及的专业领域，注意辨别多义词的一般意义和专业意义，选择国家技术标准或权威专著中采用的术语，否则就会出现词不达意的现象，违背译文的准确性原则。例如，body 的基本含义为"身体"，但专业背景不同时，含义会发生变化。

1. The sun is the central body of the solar system.

太阳是太阳系的中心天体。（天文学）

2. The speed of a plane is associated with the shape of its <u>body</u> and wings.

飞机的速度与机身及机翼的形状有关。（航空）

3. The <u>body</u> needs special exercise in a space ship to suit the weightless conditions.

人体需要在宇宙飞船中进行专门训练才能适应失重状态。（航天）

4. The falling <u>body</u> accelerates at a rate of 32 ft/s.

自由落体以 32 英尺/秒（9.75 米/秒）的速度加速落下。（物理）

再以"operation"一词为例：

1. The patient has never felt better after his bypass <u>operation</u>.

这位病人在搭桥手术后感觉非常好。（医学）

2. The Army is no longer able to conduct large-scale, division-and-corps-level combat <u>operation</u>.

陆军无法开展大规模的师团级作战行动。（军事）

3. Multiplication is a mathematical <u>operation</u> of finding the result of two or more numbers by repeated numbers' additions.

乘法是一种数学运算，通过对两个或数个数字的重复相加来求结果。（数学）

4. Dumping does not include the discharge of wastes arising from the normal <u>operation</u> of vessels, aircraft, or other vehicles and facilities.

"倾倒"不包括租用船舶、航空器及其他载运工具和设施正常操作产生的废弃物的排放。（机械工程）

5. As an aviation <u>operations</u> specialist, you will be responsible for scheduling and dispatching tactical aircraft missions and making one of the largest fleets of aircraft in the world run safely and efficiently.

作为一名航空运营专家，你将负责调度战术飞机任务，并使世界上最大的飞机编队之一安全高效地运行。（航空管理）

显然，同一个词在不同专业背景下可被译为不同的词。在此种情况下，译者应利用科学常识确定其所在文章的学科和专业，再查阅相关的专业词典，最终确定合适的词义。

（四）根据英汉搭配习惯选择词义

英语的一词多义也体现在词与词的搭配上。不同的搭配方式，会产生不同的词义。这时应当以词组为单位进行理解和翻译，不能逐词翻译。以形容词 large 为例，其本义为"大的，大量的，大号的，广泛的"等，但如果把 large 放在特定的搭配中，词义就有所不同，如表 2.1 所列。

表 2.1

术　语	译　文
large pressure	高电压
large loads	重载
large-screenTV	宽屏电视
large current	强电流
large capacity	高容量
large growing	生长快的

再如以下例子：

Increasing sophisticated computer-controlled robots are entering industry to <u>cut down</u> production costs, time and labor requirements.

日益先进的计算机控制机器人正在进入工业领域，以<u>降低</u>生产升本，<u>缩短</u>生产时间并<u>减少</u>劳动力。

动词 cut down 与三个宾语名词同时形成搭配，分别译成三个不同的词义。可见，在词义选择过程中，应当充分考虑英汉搭配习惯，选择恰当的词义。

（五）根据名词的"数"选择词义

在航空航天科技英语翻译中，区分英语可数名词（包括复数等）与不可数名词非常重要，因为科技英语中有些名词的单数形式和复数形式所表示的词义完全不同，或者复数形式除有单数形式所表达的词义外，还有新的词义，翻译时可以根据名词的单复数来选择词义，如表2.2所列。

表 2.2

单数术语	词　义	复数术语	词　义
element	元素、文件	elements	原理
main	主线、干线	mains	电源
facility	方便、条件	facilities	设备、装置、机构
charge	电荷、负荷	charges	费用

（六）根据构词规律选择词义

航空航天科技英语新词层出不穷，且很多新词都是采用构词法（word-formation）派生而来的。从形态结构来看，现代英语的构词方式主要有词缀法（affixation）和转类法（conversion）。由于航空航天科技英语术语的构词法与汉语构词法不尽相同,导致从英文中直接借用过来的专业词汇与原文传达的信息有细微差别,并造成翻译的困难。例如:sensor,transmitter,detector 三个词都是采用在动词后面加后缀-er/-or 的构词法,都可译为"传感器",但三个词所传达的信息是有差异的。sense 作为名词译为"感觉、感知";transmit 的释义为"传达、传导";detect 的词义为"察觉、探测"。在这三个词基本词义的基础上审视"传感器"这一释义,发现其并没有完全将三个词的词汇信息翻译出来,建议采用注释法翻译这组词:sensor 译为"传感器(感应式)";transmitter 译为"传感器(传导式)";detector 译为"传感器(探测式)"。通过构词规律分解词汇基本释义来理解词汇的方法,可以让译文读者更加清楚这三种传感器的特点和工作原理,也更容易理解原文。

课后练习

请将下列句子翻译为中文。

1. This is a 20 power binoculars microscope.

2. Rubber is not hard, it gives way to pressure.

3. A round surface reflector is a key unit for the solar energy device.

4. Alloys belong to a half-way house between mixtures and compounds.

5. Packaged software is developed to serve the specific needs of one user.

6. Some reactions such as those between soluble acids and bases occur very rapidly.

7. British rail engineers at its Railway Technical Center at Derby have trimmed weight and drag substantially.

8. Private aircraft are personal planes used for pleasure flying, often single-engine monoplanes with non-retractable landing gear.

9. Large missiles are typically without a rail launching system, and whether tied down or simply standing on their base are subject to wind effects.

10. The frequency with which the filter should be removed, inspected, and cleaned will be determined primarily by aircraft operating conditions.

11. Two of these gases, called butane and propane, are sometimes put into steel bottle. These gases can then be used where there is no piped supply.

12. Flight Stats Global Flight Tracker can help you to see if your flight has been delayed or cancelled and track the live position on a map.

13. For a few hours, you settle back in a deep armchair to enjoy the flight. The real escapist can watch a free film show and sip champagne on some services.

14.

原　文	译　文
solid angle	
solid body	
solid line	
solid color	
solid injection	

第 2 节　词类转换

英语和汉语分属不同的语系,在表达方式上存在着很大的差异。在词汇层面,具体体现为:英语属于静态语言,名词占比很高,作用较大;而汉语属于动态语言,动词常串联整个句子,在句中使用频率较高。在英汉互译中,词类转换常取决于译入语的表达习惯。在航空航天科技英语翻译实践中,为使译文通顺自然,更符合汉语的表达习惯,便于读者理解,在很多情形下需要进行词类转换。所谓词类转换,就是将原文中的某一词类转换成汉语中的另一词类。词类转换主要有以下四种情况。

一、转换为动词

与汉语相比,英语句子中除了谓语动词外,其他表示动作的词主要以名词、形容词、副词和介词等非动词的形式存在。而在汉语中,动词使用较多。英汉翻译时,将英语各种非动词形式转换为汉语动词形式很常见。

(一) 名词转换为动词

1. This check is a <u>control</u> on his own accuracy in observing.

这种检验就是<u>核查</u>他自己观察的准确性。(名词 control 转换为动词)

2. The <u>rotation</u> of the earth on its own axis causes the <u>change</u> from day to night.

地球绕轴自<u>转</u>,引起昼夜<u>更替</u>。(名词 rotation 和 change 分别转换为动词"转"和"更替")

3. The Committee finds that Mars is the ultimate destination for human <u>exploration</u> of the inner solar system.

委员会认为,火星是人类<u>探索</u>内太阳系的最终目的地。(名词 exploration 转换为动词)

4. This runs the risk of getting stuck at a destination without a clear

understanding of why it waschosen, which in turn can lead to <u>uncertainty</u> about when it is time to move on.

如果没有清楚了解为何选择所选的目的地,就会有卡在目的地停滞不前的风险,进而导致我们<u>无法确定</u>何时该继续进行下一步。(名词 uncertainty 转换为动词)

5. More and more scientists are visiting the Antarctic to acquire new knowledge which will help our better <u>understanding</u> of the earth as a whole.

越来越多的科学家在南极进行考察以便获得有助于我们更好地<u>了解</u>地球全貌的新知识。(名词 understanding 转换为动词)

6. Certainly, the issues at hand demand a broad and detailed understanding of the human spaceflight program——ranging from an <u>awareness</u> of the impact of galactic cosmic rays on the human body to the fact that the hook-height at NASA's Michoud Assembly Facility will only allow the manufacture of an equipment with a diameter of 33 feet.

当然,眼下的问题是需要广泛且深入地了解载人航天计划——从<u>认识</u>到银河系宇宙射线对人体的影响到美国国家航空航天局米丘德装配厂的吊钩高度只能生产直径为 33 英尺(10 米)的设备。(名词 awareness 转换为动词)

(二) 形容词转换为动词

英语中有些表示意愿、情感、态度、直觉、情绪等的形容词,在翻译时常可转换为动词。另外,英语中存在大量由动词的分词式构成的形容词,或与动词同根的形容词,这些词在翻译成汉语时,一般都转换为动词。

1. In addition, there is now a <u>burgeoning</u> commercial space industry.
除此之外,商业航天业正<u>蓬勃发展</u>。(形容词 burgeoning 转换为动词)

2. Heat is a form of energy into which all other forms are <u>convertible</u>.
热是能的一种形式,其他一切能的形式都能<u>转化</u>为热能。(形容词 convertible 转换为动词)

3. One reason the planemakers are bullish is that neither is <u>dependent</u> on the hard-hit American market.
飞机制造商看好前景的一个原因是两者都不<u>依赖</u>于遭受重创的美国市场。(形容词 dependent 转换为动词)

4. GD was well <u>aware</u> of the National Aeronautics and Space Administration (NASA) supersonic transport and fighter research efforts.
通用动力公司非常<u>了解</u>美国国家航空航天局进行的超声速运输机和战斗机研究工作。(形容词 aware 转换为动词)

5. Predictive models have generally proven unsatisfactory in accurately forecasting absolute reliability——many actual failures have been <u>attributable</u> to

causes not included in most reliability models.

在预测绝对可靠性方面,预测模型通常无法令人满意——许多实测失败是由那些大多数可靠性模型中未包括的原因<u>造成</u>的。(形容词 attributable 转换为动词)

(三) 副词转换为动词

1. The electric current flows through the circuit with the switch <u>on</u>.

<u>接通</u>开关,电流就会流过线路。(副词 on 转换为动词)

2. The two bodies are so far <u>apart</u> that the attractive force between them is negligible.

这两个物体<u>相距</u>甚远,它们之间的吸引力可以忽略不计。(副词 apart 转换为动词)

3. A human landing and extended human presence on Mars stand prominently <u>above</u> all other opportunities for exploration.

人类在火星上登陆并安营扎寨的可能性大大<u>超过</u>了其他星球。(副词 above 转换为动词)

4. It looks as if the only way Europe's proud aerospace champion can survive is by matching Boeing's <u>increasingly</u> global game.

看起来,欧洲引以为傲的宇航巨头的唯一出路似乎是与波音公司<u>日渐增强</u>的全球战略一争高下。(副词 increasingly 转换为形容词)

5. It can be prophesied that titanium materials will be used <u>increasingly</u> in aerospace and in other industrial products.

可以预言,钛材在飞机或各种飞行器上的应用将会<u>与日俱增</u>,并且在其他工业产品中的应用也会<u>扩大</u>。(副词 increasingly 转换为动词)

(四) 介词转换为动词

英语介词主要用于表示时间、地点、方式、原因、数量等,使用频率高、范围广。翻译时,英语介词常被转换为汉语动词。

1. The shaft turns <u>about</u> its axis.

轴<u>绕</u>其轴线转动。(介词 about 转换为动词)

2. We consider radar <u>of</u> great value for aviation.

我们认为雷达<u>对</u>航空很有价值。(介词 of 转换为动词)

3. The volume of a gas becomes smaller when the pressure <u>upon</u> it is increased.

当<u>作用</u>在气体上的压力增大时,气体的体积就缩小。(介词 upon 转换为动词)

4. It is clear that numerical control is the operation of machine tools <u>by</u> numbers.

很显然,数控是指机床采用数字来操控。(介词 by 转换为动词)

5. Such particles are far too tiny to be seen <u>with</u> the strongest microscope.

这种粒子实在太小,即使用最高倍的显微镜也看不见。(介词 with 转换为动词)

6. Destinations should derive from goals, and alternative architectures may be weighed <u>against</u> those goals.

目的应该源于目标,并且可以<u>根据</u>这些目标来权衡替代结构。(介词 against 转换为动词)

7. Hubble has addressed fundamental cosmic questions and explored far <u>beyond</u> the most ambitious plans of its builders.

哈勃望远镜解决了基本的宇宙问题,其探索的范围远远<u>超出</u>了其建造者的宏伟蓝图。(介词 beyond 转换为动词)

8. Alan Turing built a special-purpose electromechanical computer <u>for</u> cryptanalysis, the "Bombe", and developed a detailed specification for an "automatic computing engine", a real general-purpose stored-program computer.

艾伦·图灵建造了一台<u>用于</u>密码分析的专用机电计算机"Bombe",并为一台真正意义上的通用存储程序计算机"自动计算引擎"制定了详细的规范准则。(介词 for 转换为动词)

二、转换为名词

在英汉翻译中,原句中的动词、形容词、副词和介词等都可以根据情况,转换为汉语的名词。

(一) 动词转换为名词

在英语中,有些动词由名词派生或转化而来,且在翻译时无法找到相应的动词来代替,这时常根据汉语的表达习惯将其转换为名词;部分描述事物状态、特征的动词也常转换为汉语名词。

1. Neutrons <u>act</u> differently from protons.

中子的<u>作用</u>不同于质子。(动词 act 转换为名词)

2. Gases <u>differ</u> from solids in that the former have greater compressibility than the latter.

气体和固体的<u>区别</u>在于前者比后者有更大的可压缩性。(动词 differ 转换为名词)

3. The electronic computer is chiefly <u>characterized</u> by its accurate and rapid computations.

电子计算机的主要<u>特点</u>是计算准确而快速。(动词 characterized 转换为名词)

4. However, he was interested only in proving that the spaceship technology would <u>work</u>, not in exploiting it commercially himself.

然而,他所在意的只是证明这种航天技术的<u>可行性</u>,而不是要发掘它的商业价值。(动词 work 转换为名词)

5. The assessment <u>focused</u> on the limitations of these aircraft in likely future combat scenarios.

评估<u>重点</u>是这些战机在未来可能的战斗场景中的局限性。(动词 focused 转换为名词)

(二) 形容词转换为名词

英汉翻译中,将形容词转换为名词的情况主要有两种:形容词前面加"the"表示一类人;形容词作表语或修饰名词表示事物的特征时,其后往往加上"性""度""体"等。这里主要讨论第二种情况。

1. The atmosphere is only about eleven kilometers <u>thick</u>.

大气层大概仅有 11 千米<u>厚</u>。(形容词 thick 转换为名词)

2. In fission processes, the fission fragments are very <u>radioactive</u>.

在裂变过程中,裂变碎片的<u>放射性</u>很强。(形容词 radioactive 转换为名词)

3. The electrolytic process for producing hydrogen is not so <u>efficient</u> as the thermochemical process.

用电解法生产氢气的<u>效率</u>不像热化学法那样高。(形容词 efficient 转换为名词)

4. As an American, having NASA field a retro-reenactment of the Apollo program to get back to the moon a half-century after we sent people there the first time is <u>humiliating</u>.

半个世纪前,我们首次将人类送上月球。而今,为了重返月球,让美国国家航空航天局重新实施该计划。作为一个美国人,我认为这是一种<u>屈辱</u>。(形容词 humiliating 转换为名词)

5. Early on, this evolved into a <u>competitive</u> evaluation with the aircraft that the senior leadership within the Air Force really wanted—an air-to-ground version of the F-15 Eagle air superiority fighter known as the F-15E.

在早期,这演变成了对空军高层领导真正想要的一款飞机的<u>竞争性</u>评估——一架被称为 F-15E 的兼具对地攻击能力和空中优势的 F-15 鹰式战斗机。(形容词 competitive 转换为名词)

(三) 代词转换为名词

英语中,一般使用代词指代前文的名词,而汉语的表达习惯则倾向于重复前文的名词。将英语代词转换为汉语名词,实际上是将代词"还原"成前文的名词。

1. <u>One</u> would fall all the way down to the center of the earth without gravity.

如果没有重力,<u>人</u>会一直坠落到地心。(代词 one 转换为名词)

2. How strong an airplane is built depends on its type and purpose.

飞机的坚固程度是由飞机的类型和用途决定的。（代词 its 转换为名词）

3. Radio waves are similar to light waves except that their wave-length is much greater.

无线电波与光波相似，只不过无线电波的波长要长一些。（代词 their 转换为名词）

4. Air density decreases as the temperature goes up and it increases when it gets colder.

气温升高时，空气密度就减小；气温降低时，空气密度就增大。（两个代词 it 转换为对应的名词）

5. For nearly two thousand years, it was mistakenly believed that all heavy objects fell faster than light ones.

近两千年来，人们都错误地认为一切重量大的物体都比重量小的物体坠落得快。（代词 ones 转换为名词）

(四) 副词转换为名词

1. The only naturally occurring substance used as fuel for nuclear power is U-235.

在自然界所存在的物质中，只有铀-235 可以用于获得核燃料。（副词 naturally 转换为名词）

2. It was not until early 40's that aviation experts began to use the technique analytically.

直到 40 年代初，航空专家们才开始将这种技术用于分析工作。（副词 analytically 转换为名词）

3. Such magnetism, because it is electrically produced, is called electromagnetism.

这种磁性因为是由电产生的，所以称为电磁。（副词 electrically 转换为名词）

4. Human spaceflight is a technologically intensive activity, and during its execution new technologies are derived that have benefit to other government and commercial users of space, and to products that touch Americans daily.

载人航天是一项技术密集型活动。在实施载人航天过程中应运而生的新技术，不但对国外政府航天用户和商业航天客户大有裨益，也有利于影响美国日常生活的产品生产。（副词 technologically 转换为名词）

5. Although it is geometrically similar to the F-16's trapezoidal wing, the wing had an increased wingspan and a larger aspect ratio and wing area, and it contained a 7-percent-larger internal volume.

尽管它在<u>外形</u>上与 F-16 战机的梯形机翼相似,但其翼展更宽,展弦比和机翼面积更大,内部体积也增大了 7%。(副词 geometrically 转换为名词)

三、转换为形容词

根据上下文,也可将副词、名词转换为形容词。

(一) 副词转换为形容词

1. Gases conduct <u>best</u> at low pressures.

低压状态下,气体的导电性<u>最佳</u>。(副词 best 转换为形容词)

2. The engineer had prepared <u>meticulously</u> for his design.

工程师为这次设计做了十分<u>周密的</u>准备。(副词 meticulously 转换为形容词)

3. Robotics is so <u>closely</u> associated with cybernetics that it is sometimes mistakenly considered to be synonymous.

机器人技术与控制论<u>密切</u>相关,两者有时甚至被误认为是一回事。(副词 closely 转换为形容词)

4. These challenging initiatives must be <u>adequately</u> funded, including reserves to account for the unforeseen and unforeseeable.

这些具有挑战性的举措必须有<u>足够</u>的资金支持,包括储备金,以应对那些未被预见以及不可预见的问题。(副词 adequately 转换为形容词)

5. <u>Originally</u> designed as a lightweight air combat fighter, the aircraft was more and more often being tasked to perform ground attack missions.

该战机<u>最初</u>的设计用途是轻型空战战斗机,而后来却越来越频繁地用于执行地面攻击任务。(副词 originally 转换为形容词)

(二) 名词转换为形容词

1. It really is <u>rocket</u> science.

它真的是一项<u>高难度的</u>活动。(名词 rocket 转换为形容词)

2. The nuclear power system designed in China is of great <u>precision</u>.

中国设计的核动力系统十分<u>精确</u>。(名词 precision 转换为形容词)

3. These results exhibit more statistical fluctuations than in the previous cases, which can be attributed to the <u>abruptness</u> of the MTL curve.

与之前的情况相比,这些结果呈现出更多的统计波动,这可能是因为 MTL 曲线<u>陡峭</u>。(名词 abruptness 转换为形容词)

四、转换为副词

根据《汉语副词词典》,副词即"附于动词、形容词及其他副词的词类"。副词主要可以分为十小类：时间副词、程度副词、限度副词、情态副词、语气副词、判断副词、频次副词、关联副词、目的副词、类比副词。在航空航天科技英语翻译中，根据行文需要，其他词类也常被转换为副词，主要有以下两种情况：

（一）形容词转换为副词

修饰名词的形容词，随着名词转换为动词，相应地转换为副词。

1. A continuous increase in the temperature of a gas confined in a container will lead to a continuous increase in the internal pressure within the gas.

不断升高密闭容器内气体的温度，会使气体的内压力不断增大。（形容词 continuous 转换为副词）

2. A further word of caution regarding the selection of standard sizes of materials is necessary.

关于材料标准规格的选择，需要进一步提醒。（形容词 further 转换为副词）

（二）动词转换为副词

1. Rapid evaporation at the heating-surface tends to make the steam wet.

受热表面快速蒸发，往往会使蒸汽湿度加大。（动词 tend 转换为副词）

2. The molecules continue to stay close together, but do not continue to retain a regular fixed arrangement.

分子仍然紧密聚集在一起，但不再继续保持有规则的固定排列。（动词 continue 转换为副词）

由于英语和汉语在语言表达习惯、语言结构等方面存在差异，在英语翻译实践中需要进行词类转换，形成符合汉语习惯的表达方式。具体转换类型和转换方式多种多样，不局限于以上几类。能否转换以及转换为何种词类，译者应当结合上下文语境和对专业词汇的理解灵活运用。

课后练习

请将下列句子翻译成中文。

1. Participation by the public was extensive and the Committee made use of the reviews.

2. They paid special attention to raising the quality of magnetron and multicavity klystron.

3. Mars is unquestionably the most scientifically interesting destination in the inner solar system.

4. Astronomers have announced the possible discovery of the first known

moon outside our Solar System.

5. With goals established, questions about destinations, exploration strategies and transportation architectures can follow in a logical order.

6. The combination of technologies and innovations incorporated in its innovative design produced a significant advance in fighter performance.

7. This support included a variety of wind tunnel and spin tunnel testing, drop model tests, and simulator tests in support of flight control system development.

8. However, Congress directed that only one of these aircraft was to be funded for production as what the Air Force termed its Dual-Role Fighter (DRF).

9. In this Hubble image, the light from a far-distant galaxy is being forced by gravity to bend around the center of a galaxy cluster to form what's known as an Einstein ring.

10. Significant improvement of the cycle thermal efficiency can be achieved through the introduction of heat exchangers, with the consequent increase of the overall engine weight.

11. General Dynamics attempted to convince the Air Force that a production variant of the F-16XL was complementary to the standard F-16 and did not need to be in competition with the larger F-15.

12. During both the YF-16 Lightweight Fighter technology demonstrator program and the follow-on F-16 full-scale development programs, GD and NASA had established a highly productive working relationship.

13. High-speed ground transportation（HSGT）represents a new mode of transportation for the United States, one that is potentially superior to both airplanes and automobiles for trips between 100 and 600 miles.

14. Before the mid-20th century, practical "computers" were not machines, but people who performed mathematical computations with the aid of simple tools, such as the abacus or the slide rule.

第3节　词义引申

在航空航天科技英语汉译过程中,如何处理英汉语言的语义对应不是一件容易的事。有时会遇到在词典上找不到与原文对应的词义,或虽有对应词,但不符合汉语的表达习惯,造成词的字面意义与表达意义不一致的情况。此时,如果硬要牵强附会,照搬词典上罗列的某个释义,会使译文晦涩难懂,甚或造成误解。这时就需要对词义进行引申,采用引申翻译法。

　　所谓引申,就是根据语境、逻辑关系、搭配习惯及全句的技术含义等,不拘泥于词的字面意义或词典提供的意义,对词义做必要的调整,将该词在句子中的特定含义解释出来。比如,If any analysis is based on design solutions not completely frozen, this should be recorded in the analysis. 如果任何分析是基于不是完全固定不变的设计方法,那应在分析中对此进行记录。本句中 frozen 意为"冰封的""冻结的",这些汉语意义不符合该句的语境,若直译则只会让读者一头雾水。可以将该词转译为"固定不变的",既译出了词语的基本意义,也表达出了其在句中的特定含义。

　　从翻译的角度看,引申可分为两种:第一种是从英语本身出发,对所用词的本义作一定的调整与变动,这种情况下译者需要精准深刻理解原词的含义;第二种是原文词义比较确定,在译文表达时,译者需要做适当的变动,注重不脱离原文基础上表达的灵活性。

　　科技英语翻译中,引申法应用甚广,其类型大致可归纳为三种:专业化引申、抽象化引申及具体化引申。

一、专业化引申

　　英语科技文章中大量使用普通词汇,这些词汇在科技语境下有着不同于日常语境的专业化语义。在翻译时,应当基于其基本意义,根据所涉及专业引申出其专业化语义。例如:

　　1. The ride experts bring an understanding of the chassis, suspensions, drives, and other structural and mechanical aspects of vehicles.

　　通过机械载运器专家的介绍,大家了解了载运装置的底盘、悬挂系统、传动装置以及其他一些结构及机械特性。

　　该例中的 ride 在词典中的意思为"搭乘""驾驶""骑马"等,此处根据语境和专业含义,将其译引申译为"机械运载器";aspect 本意为"方法""层面",此处引申为"特性"。此外,该句中的其他划线词也做了类似的转译。

　　2. The major problem in manufacturing is the control of contamination and foreign materials.

　　制造时的一个主要问题是控制污染和杂质。

　　句中的 foreign materials,如果按字面直译为"外界物质",会使读者有隔物赏花之感,不禁会问"外界物质指的是什么物质",需结合语境加以专业引申,可译为"杂质",这样翻译既符合该句的专业含义,又能使读者一目了然。

　　3. If the inherent levels are found to be unsatisfactory, design modification is necessary to obtain improvement.

　　如果发现飞机固有水平不符合要求,则需要修改设计,对其加以改进。

　　在以上例句中,unsatisfactory 在词典中的意义为"不够好的""不能令人满意的",照搬词典中的意思勉强能表达出原作的意图,但是缺乏技术语气和科技文本的客观严谨性。将其转译为"不符合要求"则弥补了以上缺陷。

4. Task invisible to the user include：reading the data and map files into memory and organizing them for fast retrieval；… reacting to movements of the trackball，and house-keeping.

其不可见功能包括：将数据和导航图文件读入内存并加以处理以便快捷检索；……对移动光标做出响应，进行内务修整。

此翻译中至少有三处运用到专业化引申：task 引申译为"功能"；retrieval（提取）引申为"计算机检索"；house-keeping（家务管理）引申为"（计算机程序的）内务修整"。

二、抽象化引申

在航空航天科技英语中，有些词的字面语义比较具体或形象，这些词常被用来表示属性、概念等抽象化的意思。在翻译这些词时，若直译，则有时不合汉语表达习惯，甚至会令人费解。因此，英译汉时，译者应注意抽象与具体的转换，将该类词的词义作抽象化引申，例如：

1. America Soars Back Into Space.

美国重返太空。

Soar 这一词指的是"急速升空"这一动作，常译为"翱翔""高飞"等意。该例是一则新闻标题。标题翻译要求语言凝练、重点突出。译者将 soar back 这一短语抽象化为"重返"，意思不变，但表达上凝练简洁。

2. They reach their programmed positions within a few seconds of each other and detonate. Anything nearby is a goner.

导弹在几秒之内相继到达预定位置并起爆。附近的一切顷刻覆灭。

Goner 意为"将死之人""无法挽救的人"，若照搬原意，则文理不通，因此基于 goner 的内涵及联想意义，将其抽象译为"顷刻覆灭"，使得译文更恰如其分。

3. Superconductivity technology is now in its infancy.

超导技术正处于发展的初级阶段。

4. A single thermonuclear bomb today has the destructive force of all the bombs and other explosives of World War Ⅱ.

现在的一枚热核弹，其杀伤力相当于第二次世界大战中所有炮火的总和。

5. These systems utilized mostly chaff or similar techniques to draw the missile from its intended target，or blind its guidance.

这些系统大多采用雷达干扰或类似手段，使导弹偏离其预定目标，亦即使其制导系统失灵。

在以上例句中，infancy, bombs and other explosives, draw from, blind 等词都为表达较为具体的词，其含义分别为"婴儿期、初期""炸弹和炸药"以及"从……拔出和从……提取""弄瞎",在译文中,译者结合语境、专业技术因素和词语的基本含义,将其引申为更为抽象笼统的表达。

三、具体化引申

与汉语相比,英语重抽象理性思维,在词汇使用上常用抽象词表达具体事物和现象,这在科技英语中尤为明显。在航空航天科技英语中,有部分词汇,其字面意思非常笼统、概括和空泛,如果对应译成汉语,不但译文表达不出原作的真正含义,影响读者对整句乃至整篇文章的理解,还可能使本该严谨客观的科技文本变得似是而非、模糊不清。因此,在汉译过程中,为了明确词汇所指,应将其转换为具体表达。如:

1. For anti-aircraft and anti-missile <u>applications</u>, considerable importance must be attributed to an unmanned fully automatic gun and to a substantial reserve of ready use ammunition which can be fired without human <u>intervention</u>.

在防空和反导<u>战斗</u>中,必须特别注重无人<u>操作</u>的全自动火炮以及大量的<u>随时</u>可发射的现成弹药储备。

该例句中,application 原意为"应用""运用""应用程序"等,若要根据词典中的意思翻译成"在防控和反导应用中",读者可能会不禁疑惑"应用的是什么",以至于影响对于后半句的理解。本译文将其具体译为"战斗"则能使语义表达明确,易于读者理解。同样地,intervention 也做了相应的处理。

2. Radar waves go through clouds or fog quite <u>well</u>, whereas light waves do not.

雷达波能<u>轻易</u>穿云破雾,光波则不能。

3. The planes may be in commercial use <u>as early as the end of this decade or the beginning of the next.</u>

<u>早在 80 年代末或 90 年代初</u>,这类飞机或已投入商业运营。

4. Cold rolling enables the operators to <u>produce</u> rolls of great accuracy and uniformity, and with a better surface finish.

冷轧使操作者能<u>轧出</u>精确度和均匀性很高、表面粗糙度较好的卷材。

5. Also, there are <u>active</u> astronauts：Jon McBride, pilot of Challenger on a 1984 mission…

另外,还有一些<u>现役</u>航天员：乔恩·麦克布赖德(执行 1984 年飞行任务的"挑战者号"飞行员)……

在以上例子中,well 作为副词在词典中的意思为"好""很",produce 意为"生产""制造"以及"出产",active 的意思为"活跃的""积极的",译者根据具体的语境将这些笼统的含义具体引申为"容易地""轧出"和"现役",并且根据语境将例 3 中的时间具体译出,在不背离词语根本意思的基础上清晰地反映了原文要表达的含义,符合专业技术语体的特征,言简意赅。

四、词义引申的原则

　　航空航天类科技文章具有很强的科学性、严谨性和逻辑性,词与词之间,段与段之间,各部分与文章整体之间是相互依存的。一般来说,需要引申的词语多见于单个的词、词组或短语,引申的背景又多是与文章主题相关的概念。恰当的引申可以弥补词典释义的局限和不足,但是灵活地运用以上三种引申方法并非易事。无论使用何种引申方式,在实际翻译中需考虑汉语的表达习惯,掌握一定的航空航天领域专业背景知识。请看以下几个例子:

　　1. Colors can give more force to the form of the product.

　　原译　颜色能给予产品外形更多的力量。

　　改译　色彩能为产品外观增添美感。(利用引申法,注意汉语表达习惯)

　　2. Professor Ronald Bracewell, a leading American radio-astronomer, argued in *Nature* that such a superior civilization, on a visit to our own solar system, may have left an automatic messenger behind to await the possible awakening of an advanced civilization.

　　原译　美国的一位第一流射电天文学家罗纳德·布莱斯维尔教授在《自然》杂志上提出论据说这样的一个高级文明,在访问我们的这个太阳系时,可能留下了一个自动化的信使,以等待一个高级文明的可能的觉醒。

　　改译　美国的一位第一流射电天文学家罗纳德·布莱斯维尔教授在《自然》杂志上写道:具有如此高级文明的智慧生命,在访问我们的这个太阳系时,可能留下了一个自动通信装置,以等待某一高级文明社会的可能的觉醒。(利用引申法,注意汉语词语的搭配)

　　3. Beyond our own galaxy—the whirlpool of stars and cosmic dust of which our sun is an out-of-town member, lying on one of the remoter spiral arms—are other galaxies.

　　原译　在我们的银河系之外,还有其他的星系。我们的银河系是由一个恒星和宇宙尘埃组成的漩涡星系,我们的太阳系并非银河系的成员,而是位于一个比较遥远的悬臂上。

　　改译　在我们的银河系之外,还有其他的星系。我们的银河系是由一个恒星和宇宙尘埃组成的漩涡星系,我们的太阳系并不在这个旋涡星系的中心,而是位于一个比较遥远的悬臂上。(利用引申法,注意科学常识和专业知识)

　　从以上翻译例句中可以看出译者不仅要具有较高的中英文水平,掌握一定的专业知识,还要秉持精益求精的态度,这样才能在翻译实践中体现航空航天领域文献的科学、准确和严谨。

课后练习

请将下列句子翻译成中文,注意引申词义。

1. The temperature is well above the set value.

2. Rubber is not hard, it gives way to pressure.

3. The pupil of the eye responds to the changes of light intensity.

4. Atoms are much too small to be seen even through the most powerful microscope.

5. An electromagnet weighing about a ton can hold a piece of steel weighing 7or 8 tons.

6. Most comets are extremely faint objects, far below the limit of the unaided eye.

7. For generations, coal and oil have been regarded as the chief energy sources used to transport men from place to place.

8. We are at the dawn of the space age, and, it seems, yesterday's fantasies will be tomorrow's realities.

9. A modern frigate or destroyer could well use up all her surface-to-air missiles in the first hours of a serious large-scale encounter.

10. The conclusion of this test was that the missiles can be made to fail in its mission with the 99% confidence level, provided the first intercept occurs in time.

第4节　增译和省译

航空航天科技英语文体正式客观,逻辑严密,翻译时应力求如实体现原作的文体风格。但由于英汉两种语言分属于不同的语系,在语言组织方式上,英语属于形合语言,汉语属于意合语言,两者在行文逻辑和词汇使用上都存在较大差异。在汉译过程中,译者往往需要根据汉语的表达习惯采用增译法或省译法,使译文更加通顺、流畅、自然。

一、词的增译

增译法即在翻译时增加一些原文虽无而译文表达又不可或缺的内容。这是航空航天科技英语汉译时经常采用的方法,如:A ring of debris composed of dust, rock, and ice surrounds the Sun-like star called HBC 672, but it is too small and distant for even Hubble to see—at least directly. 由尘埃、岩石和冰组成的碎片环围绕着这

颗名为 HBC 672 的类太阳恒星,但它体积太小,<u>距离遥远</u>,即使用<u>哈勃望远镜</u>也无法看到——至少是无法直接观测到。该句翻译中,在 small 和 distant 前增译名词"体积"和"距离",Hubble 后增添"望远镜",这些增词可以使原文表达更清晰,便于读者理解,同时也体现了科技文本的专业性和严谨性。

译者在增词时要遵循一定的原则,首先要理解原句的内容和结构,并结合专业技术术语的表达规范,用符合汉语表达习惯的语言完整、准确、流畅地将原文的意思表达出来。

根据增添的词类不同,增译法可以分为以下十种类型。

(一) 增译名词

航空航天科技英语中,使用抽象词汇和普通词汇来表示专业含义的现象非常普遍,直译这些词汇,很难表达出其具体和明确的含义。因此在汉译时需要增添如"情况""措施""现象"等词汇,使其汉语意义更加确切。另外,英语中某些词要表达的意义是隐含在基本意义之中的,翻译成汉语时,应将其隐含的意思明确表达出来,这样译文才更加清晰易懂,更符合汉语的表达习惯。例如:

1. Atomic cells are <u>small</u> and very <u>light</u>, as compared to ordinary dry ones.

与普通干电池相比,原子电池<u>体积小</u>且<u>重量轻</u>。(在 small 前增加名词"体积",在 light 前增加名词"重量")

2. Design should incorporate <u>provisions</u> to avoid any accidental collapse of a bed.

在设计中要纳入各种<u>预备措施</u>以避免床层的意外崩塌。(名词 provision 后增译名词"措施")

3. From the <u>evaporation</u> of water, people know that liquid can turn into gases under certain conditions.

从水的<u>蒸发现象</u>,人们得知液体在一定的条件下能转化为气体。(名词 evaporation 后增译名词"现象")

4. Gravity is a strange force. When you slip on something, you never go up into the air, but instead you always <u>fall down</u>.

地心引力是一种奇异的力量。你滑倒时,绝不会跌向空中,而是相反,总是<u>倒向地面</u>。(动词短语 fall down 后增译名词"地面")

5. Combine digital technology with advanced software, <u>smaller</u> and more <u>powerful</u> microprocessors, and exponential growth in <u>fiber</u> and wireless bandwidth, and you get something far more powerful-seamless, universal connectivity.

把数字技术与先进的软件、<u>体积更小</u>、<u>功能更强大</u>的微型处理器以及加速增长的光纤和无线带宽相结合,就会获得功能更强大的无缝隙全方位的连接。

翻译这句时,需要根据原文所要表达的意思,在形容词前增译虽无其词但有其义的名词"体积"和"功能";还要分析专业技术的含义,在 fiber "纤维"前增加"光"字,译成"光纤"。

(二) 增译动词

根据语义和表达的需要可以在英语名词前后增译动词,使译文更符合汉语表达习惯。例如:

1. The next stage of space travel is a space station.

宇宙飞行的下一步是建立空间站。(在 space station 前面增加动词"建立")

2. These differences are due to the differences in flows of energy and matter.

这些差异是由于物质和能量流动速率的差别造成的。(增加动词"造成")

3. For maximum precision, the thermal measurements are usually made electrically.

为了获得最高的精确度,通常用电来进行热测量。(在 maximum precision 前增加动词"获得")

4. Preliminary design must establish confidence that the airplane can be built on time and at the estimated cost.

初步设计必须建立信心,相信飞机可以按估算成本按时建造。(增加动词"相信")

如果例句 1 没有在名词前增译动词"建立",而是译为"宇宙飞行的下一步是空间站";例句 2 没有增译的动词"造成的",而翻译为"这些差异是由于物质和能量流动速率的差别",都会使读者感到句子生硬,意思含糊不清,难以理解。

(三) 增译形容词

根据原文所要表达的含义,结合上下文,在部分名词后适当地增译形容词,使译文能更加符合汉语习惯。例如:

1. Speed and reliability are the chief advantages of the electronic computer.

速度快、可靠性高是电子计算机的主要优点。(名词 speed 和 reliability 后分别增译了形容词"快"和"高")

2. Perhaps the most important difference between these helicopters is their power sources.

也许这些直升机之间的最重要区别在于它们的动力源不同。(名词短语 power source 后增译了形容词"不同")

3. For years now video editing has been limited by cost to the high-end professional market.

由于成本高昂,视频编辑多年来一直局限于高端专业市场。(名词 cost 后增译了形容词"高昂")

（四）增译副词

在不改变原意的基础上，有些动词前可以增译副词使译文更加符合汉语的修辞需要和表达习惯。例如：

1. The chlorine atoms eat away at the Ozone layer.

氯原子<u>逐渐</u>吞噬掉臭氧层。（增加方式副词"逐渐"）

2. All these factors play a key role in formulation of the model.

所有这些因素<u>都</u>在模型的制定中起着关键作用。（增加范围副词"都"）

3. Experiment shows no evidence of the production of explosive forces.

实验表明，<u>丝毫</u>没有产生爆炸力的迹象。（增加程度副词"丝毫"）

4. The key to the new materials is researchers' increasing ability to manipulate substances at the molecular level.

开发出新材料的关键是研究人员在分子水平上操纵物质的能力<u>不断</u>增强。（增加频度副词"不断"）

（五）增译代词

英汉两种语言在代词的使用上有较大的差异。英语中使用大量的代词来体现上下文之间的逻辑关系，而汉语中的代词则使用较少。为了弥补这种差异，在汉译过程中，译者需要适当增译代词。所谓增译代词是指通过增译"人们""有人""我们"等泛指代词或还原出代词具体的指代意义。例如：

1. Astronomers have evidence of a few other stars too, which might have black holes as companions.

天文学家也有一些其他恒星的证据，<u>它们</u>可能有黑洞作为伴星。（增译了"它们"）

2. Unlike radio-controlled aircraft, rockets were all but invulnerable to jamming or interception.

火箭不像由无线电操控的飞机，<u>它</u>几乎不受人为干扰或半路拦截的影响。（增译了"它"）

3. There are suggestions that laser beams may ultimately replace cables in telecommunications.

<u>有人</u>提出，激光光束最终可能会取代电信电缆。（增译了"有人"）

4. With the popularity of USB peripherals, it is believed USB interfaces would become a standard feature for monitors.

随着 USB 外围设备的普及，<u>人们</u>相信 USB 接口会成为显示器的标准部件。（增译了"人们"）

(六) 增译量词

英语中的量词是很有限的,表示数量概念时往往是数词或不定冠词(a/an)与可数名词直接连用;而汉语却习惯于根据事物的形态、特征或材料,用不同的量词来表示不同事物数量的概念。因此,翻译时应根据汉语的表达习惯增加适当的量词。例如:

1. Simple in principle, the experiment led to a scientific revolution with far-reaching consequences.

这个实验虽然原理简单,却带来了一场影响深远的科学革命。(增译量词"场")

2. This machine has two settings, fast and slow.

这台机器的速度有快、慢两个挡。(增译量词"台"和"个")

3. A Boeing spokeswoman said the company offered assistance as needed during the maintenance procedure and the plane has returned to service.

波音发言人说,由于公司在飞机的维护过程中提供了必要的协助,这架飞机已经恢复运营。(增译量词"架")

4. We had four spacecraft, two around Mars, two on the surface, an amazing accomplishment.

我们共计有4个航天器,其中2个绕火星飞行,2个着陆于火星表面,成就令人惊叹。(增译量词"个")

(七) 增译表示名词复数的词

英语名词的复数一般通过复数形式表示,即加后缀"-(e)s",而汉语则靠加数词或者其他词的方式表示。在汉译时,可根据具体情况增译适当的表示复数概念的词,如"们""各种""种种""许多""大量""几个""一些"等。例如:

1. Air is a mixture of gases.

空气是多种气体的混合物。(增译"多种")

2. The mechanism of polymer degradation by ultraviolet radiations is not agreed upon by experts.

关于紫外线辐射使聚合物降解的机理,专家们是有分歧的。(增译"们")

3. For reasons the alternating current is more widely used than the direct current.

由于种种原因,交流电比直流电用得更为广泛。(增译"种种")

4. The Galileo mission is now considered to be a huge success, although there were setbacks at the beginning.

现在看来伽利略号探测器的任务完成得相当出色,虽然最初也遇到过一些挫折。(增译"一些")

(八) 增译时态词

英语的时态通过动词词形的变化来体现,而汉语中动作发生的时间和状态则通过与时间相关的副词、助词等来表达,因此在汉译过程中需要增译时态词。在翻译英语的进行时态时,往往增译"正在""在""不断""着"等词语;翻译将来时态时,增译"将""要""便""会"等;翻译过去时态时,增译"曾经""当时""以前""过去"等;翻译英语的完成时态时,增译"已经""历来""了"等。例如:

1. Rockets have found application for the exploration of the universe.

火箭已经被用来探索宇宙。(用"已经"表示完成时态)

2. If the resistance of the circuit is high, the current will decrease rapidly, but a high induced e. m. f will result.

如果电路的电阻高,电流就会迅速减少,但将产生强的感应电动势。(用"就会"和"将"表示将来时态)

3. There was expected to be a radial electrostatic field near the Earth's surface.

当时在地球表面附近兴许有一径向电场。(用"当时"表示过去时态)

4. The shell parts of reactor pressure vessels have been fabricated with formed plates welded together.

反应堆压力容器的壳体,历来是用成型钢板焊接而成的。(用"历来"表示完成时态)

5. The chemist is making an analysis of the poison.

化学家正在分析毒药。(用"正在"表示进行时态)

6. But that gas escapes from Earth into space shortly after the gas's formation.

但这些气体产生后立刻就逃离地球进入太空了。(用"了"表示完成时态)

(九) 增译概括性的词

所谓概括性词语,就是用来对所罗列的事物进行总结概括的词语,比如"两种""三类""双方""等等""种种"等。英语在列举事实时很少用概括性的词语,而汉语则恰恰相反。因此,在翻译时,为了使译文更加清晰明了,需要根据实际情况增添一些概括性的词语。例如:

1. Basic design encompasses both concept design and preliminary design.

基本设计包括概念设计和初步设计两个部分。(增译"两个部分")

2. The advantages of the recently developed composite materials are energy saving, performance efficient, corrosion resistant, long service time, and without environmental pollution.

最新开发的复合材料具有节能、性能好、抗腐蚀、寿命长和无污染五大优势。(增译"等五大优势")

3. The vapor pressure changes with the temperature, the pressure, and the kind of liquid.

蒸汽压力随温度、压力和液体类型<u>这三种因素</u>的变化而变化。(增译"这三种因素")

4. Radiant, electrical and chemical energies can all be turned into heat.

辐射能、电能和化学能<u>等能量</u>能都可以转化为热能。(增译"等能量")

5. The frequency, wave length and speed of sound are closely related.

频率、波长和声速<u>三者</u>密切相关。(增译"三者")

(十) 增译连接词

英语中不少逻辑关系是通过语义关系和语法结构等来体现的,而汉语则是通过关联词。因此,在汉译过程中,为了使译文更符合汉语的表达习惯,可以适当地在译文中增加一些表示原因、条件、目的、转折、结果、让步、假设等逻辑关系的连接词。例如:

1. Ice and water consist of the same substance in different forms.

冰和水由相同的物质构成,<u>但</u>形态不同。(增译转折连词"但")

2. Heat from the sun stirs up the atmosphere, generating winds.

太阳发出的热能搅动大气,<u>于是</u>产生了风。(增译连接词"于是")

3. Other liquids being too light, a barometer uses mercury.

由于其他液体太轻,<u>因此</u>,气压表采用水银。(增译连接词"因此")

4. There are a number of devices that give an electric current when light falls on them, without any chemical changes taking place.

有许多装置在受到光线照射时会产生电流,<u>而</u>并不发生任何化学变化。(增译连接词"而")

5. As the nature of the soil often varies considerably on the same construction site, the capacity of the soil to support loads also varies.

<u>即便</u>在同一施工场地,由于土壤的性质有很大差异,土壤的承载力也不相同。(增译连接词"即便")

课后练习

请将下列句子译为中文。

1. Crimp hinge ends after pin installation to retain pin.

2. Transistors can make previously large equipment much smaller.

3. If the reaction took hours, and not seconds, the fuel costs would be prohibitive.

4. Our sun is just an ordinary, average-sized, yellow star, near the inner edge of one of the spiral arms.

5. In the absence of sufficient knowledge confusion exists in classifying a specimen as a butterfly or a moth.

6. The rise in technology, particularly for information and communication, is radically transforming lifestyles.

7. There are three basic ways of producing or storing electricity: by generator, by battery, or by accumulator.

8. When the ends of a copper wire are joined to a device called an electric cell a steady stream of electricity flows through the wire.

9. The object of the present invention is to improve the power, applicability, consumption and reduce the cost of the internal combustion engine.

10. There is no single definition of the term, but rather a combination of elements—new or upgradedtrack, rolling stock, operating practices—that lead to high-speed rail operations.

11. It might seem countering intuitive, but signs can actually slow us down. You can't resist stopping to read them. So Atlanta airport has as few as possible. But it is not just about the actual speed at which we move. It is also about our perception.

二、词的省译

省译法是与增译法相对应的一种翻译方法,是在完整传达原文信息的前提下,删去不符合汉语思维习惯和行文习惯的词或短语。严格来说,翻译航空航天科技文章时的原文信息是不允许有任何删节的,但由于英汉两种语言表达方式的不同,英语句子中有些词如果硬是要译成汉语,反而会使得译文晦涩难懂;如果采用省译策略,则会使译文更加通顺易懂。在科技英语翻译中,省译主要包括以下七种情况。

(一) 省译冠词

英语中有冠词,而汉语中没有冠词,汉译时常常可以省去不译。例如:

1. Much of the sunlight is absorbed by the oceans.

大部分阳光被海洋吸收。(省译定冠词 the)

2. A flight in an airplane is no longer such an adventure as it used to be.

坐飞机不再像过去那样是一种冒险了。(省译不定冠词 a 和 an)

3. The ship landed in a medley of noises. There was the far-off hiss of the atmosphere cutting and sliding past the metal of the ship.

星舰在各种混杂的噪声中降落——远方传来金属舰身切入大气层时摩擦出的嘶

嘶声。(省译不定冠词 a)

4. Although the type of disk that gives rise to the shadowy feature is common around young stars, the combination of <u>an</u> edge-on view and the surrounding nebula is rare.

尽管在新生恒星周围形成阴影特征的圆盘类型很常见,但边缘视图和周围星云的组合却很少见。(省译不定冠词 an)

5. Assuming <u>a</u> favorable decision for entering full-scale development, the detail design phase begins in which the actual pieces to be fabricated are designed.

假设顺利做出了全面开发的决定,那么详细设计阶段便开始启动。在此阶段需设计出要制造的实体零件。(省译不定冠词 a)

(二) 省译代词

英语和汉语在代词的使用上有很大差别,英语较多使用代词以避免名词重复,而汉语则倾向于少用代词以求行文简洁。在航空航天科技英语的汉译中,对于没有特殊含义的代词可以省略不译,其中常被省略的有指示代词、人称代词、物主代词,以及定语从句中的关系代词等,例如:

1. The volume of the sun is about 1300000 times <u>that</u> of the earth.

太阳的体积约为地球的一百三十万倍。(省译指示代词 that)

2. If <u>you</u> reduce the volume which a gas occupies to one third, then the pressure increases three times.

如果把气体所占的体积压缩到 1/3,那么压力便会增至 3 倍。(省译人称代词 you)

3. If <u>one</u> looks at the sky on a clear, moonless night, the brightest objects <u>one</u> sees are likely to be the planets Venus, Mars, Jupiter, and Saturn.

如果在一个清朗、无月亮的夜晚仰望星空,能看到的最亮星体可能是金星、火星、木星和土星等行星。(省译人称代词 one)

4. To be sure, he had traveled previously only as far as Synnax's only satellite in order to get the data on the mechanics of meteor driftage which he needed for <u>his</u> dissertation.

其实,他以前最远只到过辛纳克斯唯一的卫星上,为撰写博士论文搜集有关陨石漂移的力学资料。(省译物主代词 his)

5. If one drops two bodies <u>that</u> don't have much air resistance, such as two different lead weights, they fall at the same rate.

如果抛下两个空气阻力不大的物体,如两个重量不同的铅锤,它们会以同样的速度坠落。(省译关系代词 that)

（三）省译介词

大量使用介词是英语的又一大特点。英语中常用介词来表达时间、地点、方位等，在汉语中此类无实际意义的介词通常可以省译。例如：

1. When a substance changes only in state or in form, it is a physic change.

当物质只发生状态或形式变化时，是物理变化。（省译介词 in）

2. That was in the 1980s, when flights were not crowded. The airline did not even charge him for the extra luggage.

那是 20 世纪 80 年代，航班并未人满为患，航空公司甚至不收行李超重费。（省译介词 in）

3. China will lift jet fuel surcharges for the second time in a year on domestic flights to help air carriers cope with the impact of soaring oil prices.

为了缓解油价上涨对航空公司造成的压力，中国在一年内将第二次上调国内航线的燃油附加费。（省译介词 on）

4. The Solar Dynamics Observatory is now NASA's best eye on the sun, with a resolution far-exceeding any previous telescope.

太阳动态观测台是美国国家航空航天局现有观测太阳最好的"眼睛"，它的分辨率远超之前的任何望远镜。（省译介词 with）

5. They want to know how much money could be saved by cancelling parts of the shuttle's successor.

他们想搞清楚取消生产新一代航天飞机的部件会节省多少资金。（省译介词 by）

6. Air France even believes it will quickly be able to make money by buying Alitalia, one of the weakest airlines in Europe.

法国航空甚至认为，收购欧洲实力最弱的意大利航空会很快实现盈利。（省译介词 in）

（四）省译连词

英语作为一种形合语言，常使用一些连词来体现上下文的逻辑关系，而汉语则是通过语义来连接句子。在翻译时，英语原文中的连词常常可以省译。例如：

1. A gas becomes hotter if it is compressed.

气体受压缩，温度就升高。（省译连词 if）

2. To every action there is an opposite and equal reaction.

每个作用力都有一个大小相同、方向相反的反作用力。（省译连词 and）

3. The first rocket reaches the end of its flight and falls to earth.

第一节火箭完成行程，坠落地面。（省译连词 and）

4. Unique what makes the space shuttle is that it takes off like a rocket but lands like an airplane.

航天飞机的独特之处是,起飞时像火箭,降落时像飞机。(省译连词 but)

5. The altitude determination of spacecraft has always been a hot problem in spaceflight and aero aviation.

航天器的高度判定一直是航天航空领域的一大热点问题。(省译连词 and)

6. China has expended a great deal of its rare resources of talent and time, and has taken enormous risks to carry out this project.

中国花费了大量宝贵的人力与时间,冒着巨大的风险实施了这一计划。(省译连词 and)

(五) 省译动词

英语句子通常由动词作谓语,而汉语句子中除了动词作谓语,形容词、名词或词组都可以作谓语,因此翻译时可以省译原文中的动词。通常英语中的某些系动词或行为动词会省译。例如:

1. Lubricants of low viscosity also exhibit a low temperature dependence of viscosity.

低黏度润滑剂的黏度对温度的依赖性也低。(省译动词 exhibit)

2. In this aircraft simulation context, uniform rectilinear movements are the norm, and variation from these occur in a gradual way.

在这项飞行模拟实验过程中,匀速直线运动是常态,但也有一些渐进的变化。(省译动词 occur)

3. The technical and economic analyses show that China should not develop the space shuttle as the manned spacecraft.

从技术和经济角度分析,中国不宜将航天飞机作为载人航天器来发展。(省译动词 show)

4. The lookout also will provide 360 degree views of the space station, allowing the crew to see directly outside during robotic work.

有了瞭望台,航天员还能以 360 度的视角观察国际空间站,直接看到机器人在外面工作的情况。(省译动词 provide)

5. If the surface is smooth, reflection is regular; if the surface is rough, the reflection is diffuse.

如果表面光滑,则反射规则;如果表面不平,则反射发散。(省译系动词 is)

(六) there be 句型的省译

There be 句型是英语中的常用句型,其意思是"有",表示"人或事物的存在"或

"某地有某物"。汉译科技英语中 there be 句型的方法有多种,其中常见的一种就是省译,即不把"有"翻译出来。例如:

1. The weather was sunny, and there was little wind obvious advantages for a plane so light and dependent on the sun.

天气阳光明媚,风很小,这对一架如此轻盈的太阳能飞机来说非常有利。(省译 there was)

2. However, after the Challenger shuttle was lost in an explosion in 1986, there was debate about what to do.

但是,1986 年挑战者号航天飞机爆炸失事后,人们对下一步该怎么办争论不休。(省译 there was)

3. There was a rise in temperature, and as a result, the component failed.

温度升高,导致元件失效。(省译 there was)

4. If there is little cooling water available, a diesel engine makes an excellent prime mover for generation below 10000 kVA.

如果可用的冷却水很少,柴油发动机为发电量低于 10000 千伏安的设备提供极好原动力。(省译 there is)

5. As a moderate there is a choice between normal water, heavy water and graphite.

可以选择普通水、重水或石墨作为减速剂。(省译 there is)

(七) 省译重复出现的词汇或近义词

在航空航天科技英语汉译中,为使译文简洁,直截了当,常省略重复出现的词汇或者近义词。

1. All the measurements must be done with the greatest accuracy and exactness.

所有测量都必须无比精确。

2. Throughout the world come into use the same signs and symbols of mathematics.

全世界都采用相同的数学符号。

在例 1 和例 2 中,accuracy 和 exactness,sign 和 symbol 为同义词,在翻译时只需译一个词义,避免重复。

3. So as a team, as a sailing team we really see firsthand the impact of ocean plastic and marine debris on the environment.

因此,作为一个团队,一个帆船小组,我们确实目睹了海洋里的塑料和碎片对环境的影响。

ocean 和 marine 为近义词,均为"海洋"的意思,若直译原文则为"海洋里的塑料和海洋里的碎片”,这样表达显得啰嗦累赘,省译效果更好。

4. In the theory of relativity there is no unique absolute time, but instead each individual has his own personal measure of time that depends on <u>where he is</u> and <u>how he is moving</u>.

在相对论中,并没有独一无二的绝对时间;相反地,每个人都有他自己对时间的判定,这取决于他所处的位置以及运动的方式。(省译第二个 he)

课后练习

请将下列句子译为中文。

1. Such an engine is called an internal combustion engine.

2. The inner and outer rings are graded and stored according to size.

3. Temperatures may reach 700℃ and surface oxidation becomes critical.

4. It takes time before ozone-depleting chemicals can be totally eliminated.

5. Vacuum breakers fail if leaks permit air to enter the interrupting chamber.

6. One material can be distinguished from another by their physical properties.

7. If the atomic nucleus of a heavy element is split, a large amount of energy is released.

8. From then on, as electricity was consumed, the meter would either reverse itself or stop entirely.

9. The melting point of a substance is the temperature at which the change of state from solid to liquid occurs.

10. An understanding of the entire design sequence is essential to anyone seeking to develop a basic design.

11. Whenever a current flows through a resistance, a potential difference exists at the two ends of the resistance.

12. Each arm would be made of a plastic film that could be rolled out onto the surface of the moon, either by robots or by astronauts.

13. At the time people had lately become infected with a powerful desire to understand the Earth to determine how old it was, and how massive, where it hung in space, and how it had come to be.

第 5 节　专业术语翻译

本章开篇提到航空航天科技英语的词汇主要分为三类:专业术语词汇、半专业术语词汇和普通词汇。本节重点讨论专业术语的翻译。航空航天科技术语指的是与航空航天领域相关的专业名词（包括名词化的词）。航空航天作为新兴的科技领域，同其他学科（如物理、化学、地理、生物、天文、通信、计算机等）领域也密切相关。在词形上，该类词汇的特点是多用缩写词、缩合词、剪切词和合成词，这也在一定程度上决定了相应术语的翻译方法。由于航空航天技术迅猛发展，新术语不断涌现，能灵活翻译专业术语成为译者不可或缺的技能之一。当今常用的术语翻译方法主要有以下八种。

一、直译法

直译，即按照词汇的字面意思逐词翻译，是航空航天科技术语翻译中最常用的方法。这种译法能传达原文意义，体现原文风格，便于读者理解。故译者在翻译术语时，可先考虑直译，例如：

error rate 出错率

orbital module 轨道舱

ablation rate 烧灼速度

attitude stability 姿态稳定性

celestial guidance 天体制导

estimated off-blocks time 估计无堵塞时间

landing area 着陆区

delay line 延迟线

airline network 航线网

equatorial orbit 赤道轨道

communication satellite 通信卫星

由以上例子可以看出，直译适用于英汉两种语言中有对应项的术语以及字面意义即可传达该术语本意的术语。这一方法要求译者有较丰富的专业背景知识，而不是机械地照搬词典。

二、意译法

有些科技术语，若仅仅翻译其字面意义，很容易让人不知所云，这时候就需要在忠实于原词汇基本意思的基础上进行扩展和引申，这就是意译。意译帮助译文摆脱原文的结构束缚，使译文符合汉语规范，例如：

falling leaf 飘降轨迹

free flow 未扰动流

clear to taxi 获准滑行

airport terminal 机场航站楼

bulk inclusive tour 往返全包式大规模转运

cabin pressure emergency relief valve 座舱压力应急安全活门

fail-safe load 安全载荷

cabin baggage 随身行李

escape tower 应急救生系统

blocked-off charter 全包航班

在以上翻译示例中，"cabin baggage"指的是旅客在乘坐飞机时被允许带上飞机的行李。若按照其字面意义直译为"机舱行李"，就会令人不知所云，这时应结合乘机的相关知识，将其意译为"随身行李"，这样既能传达原意，也使读者易于接受和理解，符合科技文本的准确性。再如"airport terminal"中的"terminal"意为"终端"，

"终点",若将其直译为"机场终点"会让人摸不着头脑,所以应结合机场这一背景,意译为"机场航站楼","机场候机楼"。不难看出,在运用意译时,译者不仅要掌握词汇的基本意思,还应对航空航天领域的专业背景知识和生活常识有所了解,译者只有自己弄清楚英语术语具体所指,才能将其准确、忠实地传达给读者。

三、音译法

航空航天科技术语的产生和发展与该领域的技术发展紧密相连。如今各国都在大力发展航空航天技术,大量的英语术语也随着航空航天事业的蓬勃发展持续涌入。在这种情况下,与新科技发明、新材料和新发现有关的专业术语在汉语中有时并没有确切对应的词汇,若意译,又会使表达繁冗复杂,读者难以理解。这时可以采用音译法,即根据术语的发音,选择与英语发音相似的汉字作为该英语术语的汉语译名,以体现术语所代表的技术概念。这种用于音译的汉字不再有本身的原意,只保留其语言和书写形式。该类词主要可以分为以下两类:

(一)计量单位名称

pint 品脱	gallon 加仑
ounce 盎司	pound 磅
carat 克拉	grain 格令
volt 伏特	baud 波特
dyne 达因	bit 比特

(二)新材料,新发现和新发明

Doppler 多普勒	clone 克隆
radar 雷达	sonar 声呐
quark 夸克	darlington 达林顿(复合晶体管)
nylon 尼龙	skydrol 斯开德劳尔(耐火液压油)

除了以上两类词汇采用音译外,还有部分词汇采取音意兼译的翻译方式,这种方法可以定义为在音译之后加上一个表示类别的词,或者一部分音译,一部分意译,使概念更加明确,便于读者通过汉语名称对科技术语有一个基本的了解,例如:

Hohmann orbit 霍曼轨道	Franklin antenna 富兰克林天线
Becquerel effect 贝克勒耳效应	Newtonian liquid 牛顿液体
Deri motor 戴利电动机	true Mach number 真马赫数

这类术语的翻译很好地实现了原语和目的语的对等,同时也反映了原语中的发明者、创造者的名字,也是对发明专利和发现的一种保护和尊重。

四、拆译法

科技术语的构成有两个显著的特点：一是大部分技术词汇来源于希腊语和拉丁语，二者是现代科技英语的基础；二是通过派生法构成新义。派生法指的是通过对词根加上各种前缀和后缀来构成新词，这种方法是现代科技英语中创造科技词汇的主要手段。针对派生而来的这些科技术语，常根据其特点采取拆译的方式来对其进行翻译。所谓拆译，是指按照构词的规律，将专业术语分解为前缀、词根、后缀，分别确定其含义后再拼合译出。凡是利用拉丁语、希腊语派生或合成的新词，大都采用拆译法译出。例如：

micro 意为"微型，微小的"：microelectronics 微电子学，microwave 微波；

meter 意为"计，表，仪"：barometer 气压计，spectrometer 分光仪；

semi 意为"一半的"：semiconductor 半导体；semicircle 半圆；

geo 意为"地球，地"：geobotany 地植物学，geothermy 地热；

super 意为"超级的"：supercapacitor 超级电容器，supercompressibility 超压缩性；

astro 意为"天文，宇宙，航天"：astroroket 航天火箭，astrophysics 天体物理学。

在词根词缀的基础上对该类词汇进行翻译，难度相对较小，只需明确其前缀、后缀和词根在航空航天特定领域的意义，再结合其他组成部分的含义译出即可。拆译出的术语可以反映术语的特点、性质等，易于理解，是译者容易掌握的一种专业术语翻译方法。

五、还原法

现代航空航天科技英语的发展趋势是用词简洁，因此用缩略科技术语成为一种新的现象，如用 CA 代指民用航空，ADF 代指自动测向仪等。随着科学技术的发展，缩略语的数量不断扩大。缩略语有三个显著特点：一是科技术语的缩略语多为约定俗成，非专业人员难以理解；二是存在一词多义现象，同一缩略语代表着多个词义，如缩写为 CA 的还有 California（加利福尼亚），Central America（中美洲），chartered accountant（特许会计师），certified acupuncturist（注册针灸师）等 12 种之多；三是呈现形式多样。有的缩略语全大写，有的全小写，有的兼有大写和小写，有的字母间有标点符号，有的则没有。为了准确无误地翻译出这些术语，必须把它们还原，即借助专业词典，根据完整术语进行翻译。例如：

g-suit 抗荷服　　　　　　　　　　ACP 邻道功率

BCC 基站代码　　　　　　　　　　ATI 自动发射机识别

FAA 联邦航空局　　　　　　　　　FPD 飞行计划数据

IFR navigation 仪表导航　　　　　constant RPM 恒定转速

这些航空航天术语经过还原后准确地反映了术语本意，易于理解。由于缩略术语具有极强的专业性,为专业人士熟练掌握,此类缩略词在部分航空航天专业书籍中不再译出。但面向非专业的读者,或是在和非专业人员进行交流时,建议用还原法译出,否则会造成理解问题,产生不必要的麻烦。

六、象译法

当术语的前半部分所采用的英语单位是以描写某种与技术概念有关的形象时,可以采用象译法处理。翻译时,可以将英文字母保留或者改译为字形与该形象相近的汉字,有时也将该字母或单词译成描写具体的词语。例如:

1. 使用原术语中的英文字母并加"型"或"形"字

A-bedplate A 形底座   C-hanger C 形吊具
D-valve D 形阀   M-wing M 形机翼
H-armature H 型转子   Z-section Z 形截面
O-ring O 型密封圈   V-pulley V 形槽滑轮
inverted-V engine 倒 V 形发动机   sliding T-handle T 形滑动手柄

2. 按术语英文意思译成相近或相同的汉语形状

cross bit 十字钻头   splice bar 鱼尾板
dovetail slot 燕尾槽   cuban eight 垂直横 8 字特技

3. 采用与英语字母近似的汉字进行翻译

I-beam 工字梁   T-square 丁字尺

通过以上例子可以看出,这类专业术语在词形上具有鲜明的特征,译者加以辨别即可。象译法的特点是形象生动、浅显易懂,容易被广大读者和科技工作者理解和接受。

七、形译法

航空航天科技术语中常涉及型号、牌号、商标名称及代表某种概念的字母,翻译时保留即可。这种照录原文,不必译出的方法叫作形译法。例如:

alpha rays α 射线   pH value pH 值
X-ray X 射线   Y-axis Y 轴
C-band C 波段   gamma-rays γ 射线
L-electron L 层电子   z-translation 沿法轴线性位移

象译法和形译法的区别在于前者保留的字母代表具体实物的形象,后者所保留的字母原形本身包含着特定的技术含义。形译在专业术语翻译实践中占比较小,容易辨别。

八、释译法

释译法是用解释的方法来传达原意。在采用其他翻译方法都不能达到翻译目的时,译者可以采用释译。但是释译要适度,如果译名过长或解释过多,就不符合汉语名称简略的特点。例如:

climaxing stall	极限状态失速
docking aids	保证对接的辅助设备
decayed object	从轨道上衰落的物体
dash speed	短时间内达到的最大速度
compressibility stall	压缩性气流分离
unsafe landing	在复杂气象条件下着陆
intermediate layover	飞行途中短暂停留
disabled passenger	身陷航空事故的乘客
dirty aircraft	起落架和机翼增升装置放下的飞机
catering management document	机上膳食保障服务细则

由上述例子可知,经过释译的航空航天科技术语较长,但易于理解,利于原意的传达,必要时译者可以考虑此种翻译方法。

课后练习

1. 运用直译法翻译以下术语。

space tracking and data network

polar-orbiting meteorological satellite

quick-look photographic reconnaissance satellite

radio transmission photographic reconnaissance satellite

2. 运用音译法翻译以下术语。

Prandtl number

Poisson's constant

Monte Carlo restoration

parametric Wiener filter

Mercator's plotting chart

Apollo-Soyuz test project(ASTP)

3. 运用还原法翻译以下术语。

OMS engine

NIFE accumulator

DAP rate dead band

portable LSS

NAS screw

homomorphic DPCM

4. 运用释义法翻译以下术语。

taxi post

propulsive lift

pro rata charter

propeller clearance

throat of the nozzle

profile thrust mode

第3章 航空航天科技英语句法翻译策略

航空航天科技英语属于科技题材英语。科技英语的文体风格与其他题材不同，讲究行文规范、严谨客观，用词简洁准确。其专业术语较多，句子结构复杂，逻辑性强。这些特点在句法层面主要体现在名词化结构、被动语态、无人称句、非限定性动词以及长句的使用上。

第1节 名词化结构的翻译

科技英语语法现象比较复杂，其中最显著的特征之一是广泛使用名词化结构。究其原因，科技英语要求以事实为基础来记述客观事物、反映现实的物质世界及其规律。作者在语篇构建过程中要求"见物"，而名词正是表示人或事物的词类。科技英语通过名词化结构，隐藏了深层的由动词所表达的过程和形容词所表达的特征，体现了事物化的概念抽象要求，使语义高度浓缩，结构简洁，逻辑关系清晰。

所谓名词化就是把小句变成名词或者名词词组，从而使表示过程的动词和表示属性（特征）的形容词具有了名词的特征。在本节中，为了便于探讨各种名词结构的翻译策略和技巧，我们将名词化结构的范围扩展为"简单名词化结构""复合名词化结构"和"动词或形容词衍生的名词化结构"。

简单名词化结构即名词连用，指的是由一个或多个名词修饰一个中心名词所构成的结构，如 Beijing Aerospace Control Center（北京航天飞行控制中心）、spacecraft launch site（航天器发射场）、space mission（航天任务）等。复合名词化结构是由一个中心名词和一个或多个形容词、分词、名词、副词或介词短语等前置或后置修饰语构成，如 manned spacecraft（载人飞船）、data processing system（数据处理系统）、optical tracking theodolite（光学跟踪经纬仪）等。动词或形容词衍生的名词化结构比较好理解，也是狭义上的名词化结构，是由实义动词或形容词派生的名词构成的，如 space exploration（空间探测）、hard landing（硬着陆）、confidence in flying a aircraft（驾驶飞机的信心）等。

在翻译航空航天科技英语中的上述名词化结构时，可根据具体情况，灵活地使用直译法、转译法和分译法三种翻译策略。

一、直译法，译为汉语名词词组

简单名词化结构和复合名词化结构可采用直译或顺译，直接转换为汉语对应的名词词组。例如：

1. Landing distance is largely determined by wing loading.

降落距离主要取决于翼载荷。

2. Each book presents some of Hubble's more recent and important observations within a particular topic.

每本书都介绍了哈勃望远镜在某一特定主题中的一些最近的重要观察结果。

3. Soon after, a satellite communications working group (SATCOM WG) was also formed and included experts from Inmarsat and Thales.

不久，便成立了卫星通信工作小组，组员有来自国际海事卫星组织和泰利斯的专家。

二、转译法，译为汉语的动词或形容词词类

科技英语名词化结构中，有相当一部分是以实义动词或形容词派生的名词作为中心词构成的。汉语语法规则并不限制句子中的谓语动词数量，在翻译时，可以根据汉语的表达习惯，将动词派生的名词化结构还原成汉语动宾结构或主谓结构，将形容词派生的名词化结构还原为对应的汉语形容词。例如：

1. Rockets have found application for the exploration of the universe.

火箭被用于探索宇宙。

2. The Committee finds that Mars is the ultimate destination for human exploration of the inner solar system.

委员会认为，火星是人类探索内太阳系的最终目的地。

3. The surface search was focused on the identification and recovery of any debris from the aircraft floating on the sea surface.

海面搜寻的重点是识别和回收漂浮在海面上的飞机残骸。

4. The construction of such satellites has now been realized, its realization being supported with all the achievements of modern science.

由于现代科学取得的成就，现在已经能建造这样的卫星了。

以上四例中，画线部分均为实义动词派生的名词作为中心词的名词化结构，在翻译时，将其转化为动词结构，简洁、通顺，更符合汉语的表达方式。

5. It is a success to replace steam with mercury vapor.

用水银蒸气代替水蒸气很成功。

6. Over the next few years he invented and produced many new items, including the mimeograph machine, wax wrapping paper, and improvements of the telegraph.

随后数年间，他发明并生产了许多新产品，其中包括油印机、蜡纸和改良的电报机。

7. Plants began the process of land colonization about 450 million years ago, accompanied by the <u>necessity</u> of tiny mites and other organisms that they needed to break down and recycle dead organic matter on their behalf.

大约 4.5 亿年前,植物开始移居陆地,与其为伴的还有<u>必不可少的</u>小螨虫和其他生物。植物需要它们来分解死亡的有机物质,使之再循环。

三、分译法,译为独立的分句

在短语结构规则方面,英语和汉语差异较大。汉语是一种动态语言,一句话中动辄出现几个谓语动词,以字或词为基本的句法成分,较少使用长且复杂的短语;英语是一种静态语言,少用谓语动词,在句法上的一个表现为将动词名词化,而为了将各种复杂的信息表达完整,常使用前置或后置修饰语对名词进行解释说明,形成较长且复杂的名词化结构,这种情况,可考虑将其译为独立的分句。另外,如果英语句子中有内在的逻辑关系,如名词化结构与句子其他的结构成分存在因果关系、递进关系、转折关系等,也可考虑使用此种翻译方法。例如:

1. The drift modelling was also informed by <u>the deployment of self-locating datum marker buoys (SLDMB) from aircraft and vessels</u> throughout the surface search.

在整个海面搜寻过程中,<u>飞机和船只都部署了自定位基准标志浮标</u>,这也为漂移建模提供了依据。

2. Furthermore, the scientific community that studies Mars generally agrees that its exploration could be significantly enhanced by <u>direct participation of astronaut explorers</u>.

除此之外,火星研究界普遍认为,<u>航天员直接参与</u>,能显著推动火星探索进程。

3. <u>The rotation of the earth on its own axis</u> causes the change from day to night.

<u>地球绕地轴自转</u>,昼夜更替因此产生。

以上三例中,例1和例2句子都较长,结构也相对复杂,划线部分的名词性结构也比较长,包含的信息量大,此时将它们分译成为独立的分句,符合汉语意合的特点。例 3 较短,但名词化结构"the rotation of the earth on its own axis"与之后的"the change from day to night"存在因果关系,因此可以将其译为独立的分句,做原因状语。

除此之外,以上三例的翻译,还运用了转译法,划线部分均为实义动词派生的名词化结构,在翻译时都还原为主谓结构。由此可知,在航空航天文本名词化结构翻译中,译者可根据具体情况,综合使用各翻译方法。

课后练习

请翻译下列句子,注意其中的名词化结构。

1. The rate of discovery with Hubble is simply unparalleled for any telescope in the history of astronomy.

2. Can we explore with reasonable assurance of human safety?

3. However, this modern realization of the cosmic role and significance of galaxies, including our own, took centuries to emerge.

4. Assessment of innovative value was derived by an evaluation of possible impacts research leadership may have in a specific area.

5. Originally, the second phase emphasized construction and utilization of the multifunctional airborne testbed for research and development.

6. Examination of images and recovery of items by surface vessels did not identify any items considered associated with MH370.

7. The position was completely reversed by Haber's development of the utilization of nitrogen from the air.

第 2 节　被动语态的翻译

被动语态是对事物客观状态和过程的陈述,其表述切实、客观,不带个人感情色彩,而科技英语的描述往往强调事物、现象、过程和规律的客观性,因此更多地使用被动语态。同时,被动语态强调受事者,将受事者置于句首可以达到强调客观性的目的。英国利兹大学的 John Swales 统计发现科技英语中的谓语至少有三分之一是被动语态。

为了更好地将航空航天科技英语中的被动语态翻译成汉语,首先需要了解被动语态在英汉两种语言中的使用差异。根据《现代汉语词典》的有关释义:"被动式"说明主语所表示的人或事物是被动者的语法格式。汉语的被动式有时没有形式上的标志,例如:"他选上了""麦子收割了"。有时在动词前边加助词"被",如:反动统治被推翻了。有时在动词前边加介词"被",引进主动者,如:敌人被我们歼灭了(口语里常常用"叫"或"让")。因此,汉语中被动语态并不仅仅等同于狭义的"被"字句,可根据有无被动标记分为有标记被动和无标记被动。另外,汉语中的被动标记除了"被"字外,与"被"意义相同或相似的"叫""让""挨""给"等助词也可以表示被动意义。而英语中的被动语态就相对简单很多,是通过 be 动词或 get 加上过去分词组合而成的。

与汉语相比,英语中被动语态的使用频率相当高,尤其是科技英语文体。

McEnery 和肖中华通过 Freiburg-LOB 语料库（FLOB）、LCMC、BNCdemo 和
Callhome 普通话口语转写语料库（Callhome Mandarin Chinese Transcript）这四个语
料库对英语和汉语的有标记被动进行了对比研究，发现英语中有标记被动句频率几
乎是汉语的 10 倍，且英语的被动语态统计还未包括副词插入语的被动语态。

　　形成这一显著差异的主要原因有以下几点。首先，英语中只要是及物动词几乎
都可用于被动语态，这比汉语中被动标记词通常搭配的动词范围要大得多。其次，英
文的被动语态没有明显的色彩意义，多表达中性含义，而在汉语中，被动语态（尤其是
"被"字句）的传统用法是表达否定或负面的含义，这种感情色彩可以从施事、受事、说
话人角度表现出来。随着语言的发展，汉语受西方语言的影响，有些"被"字句也可以
表示中性或褒义的感情色彩，但是仍以表示不愉快的感情色彩为主。最后，英语有过
度使用被动语态的倾向，尤其是正式文体中。在不需要说出施事者、不愿说出施事
者、无法说出施事者或是为了便于连贯上下文等情况下，英语多使用被动语态，而汉
语通常倾向于避免使用被动语态。

　　由于被动语态在英汉两种语言中使用场景、频率及含义上的差异，在处理航空航
天科技英语中被动语态的汉译时，一定要避免全部直译为被动句，这样不仅译文生
硬，翻译腔过重，还不符合汉语的表达习惯，而应该具体情况具体分析，灵活地译为汉
语中的被动语态或主动语态。

一、译为汉语的被动语态

当英语被动语态是为了以下目的时，可将其翻译为汉语的被动语态：

1. 强调受事者、动作或者过程；

2. 没有提及施事者是谁；

3. 突出重点，使上下文更为连贯。

根据汉语中有无被动标记词，可以细分为以下两种情况。

（一）译为有标记被动句

汉语中除了"被"之外，"叫""让""挨""给""受""给予""加以""为""为……所""被
……所""把""遭"等也可作为被动标记词。一般而言，若英语被动语态是为了强调受
事者或者动作本身，特别是当该动作表达的是负面或否定的含义时，可直译为汉语中
的有标记被动。如：

1. On the far side of the rings, the moon Enceladus is lit by Saturnshine, or
sunlight reflected from the planet.

　　在土星环的远端，土卫二被土星的光芒照亮，或者说被土星反射的太阳光照亮。

2. By the 1970s, the role of the navigator was being eliminated, and by the
1980s, we saw the flight engineer's job begin to disappear.

　　到 20 世纪 70 年代，导航员岗位逐渐被取消；到 20 世纪 80 年代，飞行工程师这

种职业也开始淡出视野。

3. The RAIDS device measured the thermosphere, which creates atmospheric drag on space vehicles and satellites and is affected by solar activity.

RAIDS(大气层及电离层遥距探测系统)装置测量了热大气层。这种热大气层会对飞行器和卫星造成大气阻力,并受到太阳活动的影响。

4. The feeble light from this embryonic galaxy, named SPT0615-JD, was magnified and stretched out by the intense gravity of a much closer group of galaxies.

这个名为SPT0615-JD的星系尚处于萌芽阶段,它发出的微弱光线被一个距离更近的星系群的强大引力放大和拉伸。

5. At 3,200 miles (5 150 kilometers) across, Titan is slightly larger than planet Mercury and is also perhaps the most mysterious moon in our solar system because (if for no other reason) the moon's surface is shrouded beneath an orange haze.

土卫六的直径是3200英里(5 150千米),比水星(直径)稍大。它可能是太阳系中最神秘的卫星,因为(如果没有其他原因的话)其表面为橙色薄雾所笼罩。

以上五例中,英语句子中的被动语态都被译为汉语中有标记的被动句。

(二) 译为无标记被动句

当英语被动句中没有提及施事者或者根本不存在施事者,并且该动作在汉语中可用主动形式表达被动含义时,就可以翻译为汉语中的无标记被动句。无标记被动句指的是汉语句子中没有被动标记词,以主动形式表达被动的含义。例如:

1. When the laser beam was turned off, the airplane glided to a landing.

激光束关闭后,飞机滑翔着陆。

2. Titan's surface is coated in organic molecules that form in the upper atmosphere.

土卫六的表面覆盖着上层大气中形成的有机分子。

3. In 2009, a significant space exploration goal was reached when the number of astronauts capable of living aboard the ISS increased from three to six.

2009年,可以居住在国际空间站的航天员从3人增加到6人时,这意味着一个意义非凡的太空探索目标实现了。

4. The EF (Exposed Facility) is directly exposed to space, and it is a unique facility among ISS laboratories because it enables long-term experiments in open space as well as Earth and astronomical observations.

舱外设施(EF)直接暴露在太空中,因其有助于完成长期舱外太空实验以及进行地球和其他天文观测,所以在诸多国际空间站实验室当中,该设施别具一格。

以上四例中,原句的被动语态全部都没有提及施事者,同时,它们表达的动作在翻译成汉语后,都可以用主动形态表达被动含义。这样将英语被动语态处理成汉语无标记被动句,原句受事者依然是句子主语,既突出了动作和过程,也符合汉语表达习惯。

二、译为汉语的主动语态

在不强调受事者或发生的动作时,可考虑将英语被动语态译为汉语主动语态。根据英语句子结构不同,可分为以下五种。

(一) 改变主语

当英语句子中有明显的受事者和施事者时,可将施事者作为主语,受事者作为宾语,将句子翻译为汉语的主动语态;有些句子中含有带地点状语或方式状语的介词短语,这时可将介词后的名词作为汉语句子的主语,受事者作为句子宾语,译为汉语主动句。例如:

1. Matter is transported in a circulatory system.

循环系统输送物质。

本例中,被动结构后跟着作地点状语的介词短语"in a circulatory system",原句表达含义"物质是在循环系统中被输送的",也就是"循环系统是用来输送物质的",这样将介词后的名词短语"a circulatory system"译为句子的主语,受事者"matter"译为宾语,将原句译为汉语中的主动语态。若将该句译为被动句"物质由循环系统输送",同样顺畅,也符合汉语的表达习惯。

2. Hubble's Cosmic Origins Spectrograph measured how much light was absorbed by different elements in the Leading Arm's gas.

哈勃的宇宙起源光谱仪测量了前臂气体中不同元素吸收了多少光。

3. The Advanced Prototype Development Program was formally implemented by the DOD in early 1971.

美国国防部(DOD)于1971年初正式实施了高级样机研发计划。

例2和例3中有明显的受事者和施事者,并且不是消极或否定意义的表达。在这种情况下,为了更加符合汉语表达习惯,可将施事者作为主语,受事者作为宾语转化为主动语态。

(二) 增加主语

当原句中没有提及施事者,又不强调受事者或动作时,可根据具体情况,添加描述动作的逻辑主语或泛指主语,如"人们""大家""我们"和"有人"等,译为主动句。例如:

1. Commands were stored in an on-board command queue and executed based

on system time supplied by the ISS.

研究人员将命令<u>存储</u>在主板命令队列中,并根据国际空间站提供的系统时间<u>执行该命令</u>。

2. Aeronautics <u>was still considered</u> an important area to support and explore, but starting in the early 1960s, the lion's share of the new agency's focus and funding <u>was directed</u> toward space.

人们<u>仍然认为</u>航空学是一个需要大力支持和深入探索的核心领域,但从 20 世纪 60 年代初开始,这个新机构的研究<u>重点和资金大部分都集中在</u>航天领域。

例 1 中命令被存储和执行,原句没有提及施事者,可根据句意增译逻辑主语"研究人员",更符合汉语表达习惯;例 2 含有两个被动结构,"was still considered"没有提及施事者,但一来原句动作不表达消极含义,二来没有强调受事者或动作,只是阐述这样一个事实,所以根据主句部分内容添加泛指主语"人们"翻译为汉语主动语态;"was directed"译为"重点和资金集中在",以主动形式表达被动含义。

(三) 译为无主句

科技英语使用被动语态来阐述科学事实、过程和理论的情况非常多见。当英语被动语态无法或者无须指明施事者,特别是表达观点、态度、要求、号召等情况时,可以将受事者作为宾语,翻译为汉语中的无主句。例如:

1. Research <u>is needed</u> to determine what the remote pilot can handle.

需要进行研究来确定远程飞行员能够处理什么样的情况。

2. Kibo <u>was designed and developed</u> with a view to conducting scientific research activities on orbit.

设计开发日本希望号实验舱(Kibo)是为了在轨道上开展科学研究活动。

3. During its nine missions, over 22,000 Earth images <u>were taken</u> that documented long-term study sites and dynamic events on the Earth's surface.

在其执行的 9 次任务中,共拍摄了 22 000 多张地球影像。这些图片记录了地球表面的长期研究地点以及动态事件。

以上三例中,被动结构都没有也无须指明施事者,这种情况下可将受事者转换为宾语译为汉语中的无主句。

(四) 译为判断句

判断句指用汉语中"是……的"这一句型,表达事物的状态、下定义或分类,这与科技英语中描述事物过程、性质、状态的静态被动句可以对应。例如:

1. These clouds <u>are created</u> by cold, dry air blowing over a lake and accumulating water vapor.

这些积云<u>是</u>由干燥的冷空气吹过湖面并积聚水汽<u>形成的</u>。

2. The view was acquired at a distance of approximately 209,000 miles (336, 000 kilometers) from Saturn.

这种景象是距离土星约 209 000 英里(336 000 千米)时拍摄的。

3. Disturbances seen within these sub-sections provide evidence that they have been built up by smaller galaxy clusters colliding and merging.

这些子区域中观测到的扰动证明它们是由较小的星系团碰撞和合并而形成的。

(五)"it"作形式主语的被动句翻译

科技英语中常见"it＋be＋过去分词＋从句"这种句型,其中"it"作为形式主语代替后面的从句,成为受事者。因为英语句子中没有施事者,汉译时,可以将这类句型处理成汉语无主句,也可采用前文提到的译法增添泛指主语。例如:

1. From the program's inception, it was intended that NASA technical expertise and facilities would be used by the aircraft development contractors to facilitate the implementation of advanced aeronautical technologies into their LWF designs.

从项目一开始,就计划让战机研发承包商 利用美国国家航空航天局的技术专长和设施来促进先进的航空技术应用于轻型战机设计(LWF)。

2. While it is generally thought that the pilot must have ultimate responsibility for the flight, Dr. Sheridan thought there are many tasks that the pilot may not be able to effectively perform.

虽然(大家)普遍认为飞行员必须对飞行负最终责任,但谢里登博士认为有很多飞行任务可能是飞行员无法有效执行的。

3. It is believed that the original oxygen was formed by the separation of water into hydrogen and oxygen molecules in the upper atmosphere.

人们/科学家认为(据信),最初的氧是由高层大气中的水分离为氢分子和氧分子而形成的。

例 2 中"it is generally thought"和例 3 中 "it is believed"表达的都是普遍的认知或想法,因此汉译时可根据情况,增译泛指主语,也可译为无主句;例 3 中出现被动结构"was formed"是对"最初的氧"形成过程的描述,可以译为汉语中的判断句。

"it"作形式主语的被动句在科技英语中很常见,其中很多在汉语中有约定俗成的表达模式,现将使用较多的句型及其汉译列举如下:

It is believed that…	人们相信/据信……
It is reported that…	据报道……
It is noticed that…	有人指出/人们注意到……
It is thought that…	大家/人们认为……
It is suggested that…	(有人)建议……

It is well-known that…	众所周知/大家都知道……
It has been found that…	(人们)已经发现……
It is still to be hoped that…	我们仍然希望……
It should be realized that…	(我们)应该认识到……
It is sometimes asked that…	人们有时会问……
It is universally accepted that…	人们普遍认为……
It is unanimously agreed that…	大家一致同意……

课后练习

请翻译下列句子,注意其中的被动语态。

1. Radioactivity is found on both sides of the failure plane in all cases.

2. Hubble is expected to continue operating well beyond 2020.

3. He suggested that the issue of pilot incapacitation must be thoroughly addressed.

4. A box shows a small region of the galaxy where six star clusters are highlighted in the lower left.

5. Specifically, he stated that, if the second pilot is replaced by someone at a ground station, security risk is impacted greatly.

6. Originally designed as a lightweight air combat fighter, the aircraft was more and more often being tasked to perform ground attack missions.

7. Saturn's second largest moon, Rhea (949 miles, or 1,528 kilometers, across), is bisected by the right edge of the image.

8. The outer wings and the horizontal stabilizer (as big as a smaller jet's wing) will be made of carbon-fiber composite materials.

9. Meteor spectra are commonly recorded by ground or aircraft instruments and compared to synthetic spectra to determine elemental abundances and temperatures.

10. The solar spectral irradiance (SSI) is also measured to determine how that energy is distributedamong different wavelengths and where in the atmosphere that energy is absorbed.

11. The use of nightlight data by weather forecasters is growing as the VIIRS (visible infrared imaging radiometer) instrument can be used to see clouds at night when they are lit by moonlight and lightning.

12. The flight-test program progressed at an impressive pace and was completed by January 31, 1975, with a total of 439 flight hours accomplished on

347 test sorties using the two experimental prototypes.

13. Its measurements of wind speed and direction over the ocean surface were used by agencies nationwide for weather and marine forecasting as well as for monitoring hurricanes and tropical cyclones.

14. This greatly enhanced and colorized image shows the enormous extent of the fainter，larger-scale component of Enceladus'plume，which is produced by water jets spraying through fissures at the south pole.

15. It is envisioned that a FOSS (Fiber Optic Sensing System) could collect information on how aerodynamic forces are affecting an aircraft in real-time and loop that information directly into the aircraft's control system for fuel efficiency，safety and a more comfortable ride for passengers.

16. However，it should be noted that，throughout the conversation regarding pilot incapacitation，there were several comments suggesting that some attendees felt that the issue of pilot incapacitation was being overemphasized.

第3节　无人称句的翻译

科技英语强调事物、现象、过程和规律的客观性，因此会频繁使用"无人称句"。需要注意的是，与汉语的无主句不同，英语中的无人称句主要包括以下三种句型：

1. 以"it"作为形式主语，而把真正做主语的不定式、动名词或主语从句放在谓语之后的句式。如"It is possible/necessary to do…""It is likely that…""It is a pity/shame that…""It should be noted that…"等等。科技英语明显倾向于使用"it"作形式主语的句型，形成了一定的程式化句式。

2. 带有表语从句的倒装句式。如"The fact/question is that…""Particularly noteworthy is that…"等。

3. 带有宾语从句的名词化句式。如"Results show that…""Research has proved that…""Observations imply that…"等。

在这三种英语无人称句中，第二种和第三种的翻译方法相对来说要简单一些，可基本按照原文句子的语序采用顺译法，上面的例句如"The fact/question is that…""Particularly noteworthy is that…"可翻译为"事实/问题是……""特别值得注意的是……"；"Results show that…""Research has proved that…""Observations imply that…"可翻译为"结果显示……""研究证明……""观察结果表明……"。而上述第一种"it"作形式主语的句型，其译法更多样，可根据具体情况采用顺译法、倒译法、增译法和分译法等翻译策略。

一、顺译法

顺译法指的是基本按照英语原句的语序,省掉形式主语"it"不译,将整句翻译为汉语无主句。当形式主语"it"后接被动结构(在本章第2节中详细讨论过),或者原句不强调动作的施事者时,较多地使用顺译法,译为无主句。例如:

1. At speeds up to Mach 3+, it was difficult to separate the aerodynamic from the thermal effects on the aircraft.

当速度达到3马赫以上时,很难将飞机上的空气动力效应和热效应分开。

2. It is not possible to smell the fuel, see the weather and terrain, hear the engine starting or feel the movement from a ground control center.

在地面控制中心,无法闻到燃油气味、查看天气和地形、听到发动机启动或感觉到飞行器移动。

3. He presented the possibility that, if something dire occurred, it might be possible to have these other, onboard personnel interact with automation or ground personnel.

他提出了这样一种可能性,即如果发生了紧急情况,有可能让这些其他机组人员操作自动设备或与地面人员进行交流。

4. Because a single-pilot cockpit would presumably include relatively more automated systems, it should be noted that human-human conflicts could arise, and human-machine conflicts also could arise.

由于单人驾驶舱可能会使用相对较多的自动化系统,因此需要注意,可能会出现人与人之间的冲突,也可能会出现人机冲突。

5. Given the uncertainty, it was decided that a seafloor search using the autonomous underwater vehicle (AUV) on board Ocean Shield should be performed to fully investigate the detections by looking for an aircraft debris field on the seafloor.

鉴于存在不确定性,决定利用海盾号上安装的自动水下航行器(AUV)进行海底搜索,通过在海底寻找飞机残骸区域,全面调查探测情况。

以上五例中,例1、例2和例3都是在表达客观的现实情况,描述的动作并不强调具体的施事者,因此可以使用顺译法翻译为汉语中的无主句;例4和例5中,形式主语"it"后跟的是被动结构,可翻译为汉语中的无主句。

二、倒译法

倒译法指的是颠倒原文语序进行翻译,将英语原句按照汉语表达习惯进行前后调换。使用倒译法翻译"it"作形式主语的句子时,一般做法是将位于句尾的真正主语提到句首。

当句子中真正的主语是不定式短语时,可根据具体情况考虑采用倒译法,将原句中位于句尾的真正主语挪到句首,作译文的主语。例如:

1. But it was next to impossible to conduct secret testing in the skies over the settled landscape of Ohio.

但是,在俄亥俄州居住区的上空进行秘密测试几乎是不可能的。

2. It was possible to fly a high-performance lifting-body re-entry vehicle into space and bring it back, unpowered, to a runway landing here on Earth.

驾驶一架高性能的升力体再入飞行器进入太空,在没有动力的情况下返回地球并跑道着陆是有可能完成的。

3. It will probably be a challenge to have the adequate bandwidth for communication and surveillance systems that support this real-time interaction.

为支持这种实时交互的通信和监视系统提供足够的宽带可能是一个挑战。

第二种情况,如果原句中出现了所描述动作的逻辑主语(主要指的是 It+be+adj.+for sb. to do sth. 句型),通常将形式主语"it"省略不译,将逻辑主语提前到句首作为译文的主语。例如:

4. In 1946, researchers were simply trying to see whether it was possible for an aircraft to surpass the speed of sound and return safely.

1946 年,研究人员只是想知道飞机是否有可能超过声速并安全返回。

5. Without these markings defining the runway against the lakebeds, it was difficult for pilots to tell exactly how far off the ground they were.

如果没有这些标志线将跑道和湖床区分开,飞行员就很难准确判断他们离地面有多远。

以上五例中,例1、例2和例3句子真正的主语都是不定式结构,在翻译时可将它们提前到句首作译文的主语;例4和例5中,不定式结构前都有所描述动作的逻辑主语,因此可理所当然地将逻辑主语译为汉语中的主语。

三、增译法

增译法是指在译文中增加如"人们""我们""有人"等泛指主语。当形式主语"it"后跟的是被动结构(在本章第2节中详细讨论过)或者不定式短语,并且表达的动作并不是由特定的某个人完成,而是一个泛指的行为时,可以考虑增译泛指主语。例如:

1. However, it was also noted that environmental factors could affect both ULB (underwater locator beacon)'s frequency equally.

但也有人指出,环境因素可能会对这两个水下定位信标(ULB)的频率都产生影响。

2. It is not difficult to imagine how the wingman could be helpful to a single pilot.

大家不难想象僚机飞行员能为单独驾驶的飞行员提供哪些帮助。

3. In today's high-speed jet world, it may be difficult to fully appreciate just how elusive that goal seemed to those attempting to reach it in 1946 and 1947.

如今世界已经有了高速喷气式飞机,我们可能很难完全理解,对于那些试图在 1946 年和 1947 年就达到该目标的人来说,完成这个任务是多么艰难。

4. If an asteroid richer in carbon was headed our way, it might be necessary to give it a series of gentler nudges to prevent it from breaking up.

如果一颗碳含量更高的小行星朝地球飞来,(我们)可能有必要轻轻地将其推出轨道,以防止它解体。

以上四例中,例 1 是被动结构,没有提及施事者,可增译泛指主语"有人";例 2 和例 3 是原作者想要引起读者的共鸣,描述的是一个泛指的行为,可以增译泛指主语"大家"或"我们";例 4 不定式结构要表达的是这件事要做,但并不强调需要谁去做,可以增译泛指主语,同时也可以使用第一种顺译法处理成汉语中的无主句。

四、分译法

分译法指的就是将原句位于句尾的真正主语先译为独立的小短句,把形式主语"it"及其之后的谓语部分译为"这(是)……",从而将原句断为几个短句。分译法一般用来强调原句中真正的主语部分,尤其是当原句中真正的主语过长,结构复杂时,可采用该法。例如:

1. It is important that the proportion of air and fuel drawn into the cylinder are correct.

进入气缸中的空气和燃料的比例要合适,这一点很重要。

2. It is not unheard of for researchers and technicians to work through the night to find a solution to a potential flight-stopping problem.

研究人员和技术人员通宵达旦地工作,想要解决可能造成航班停飞的问题,这并非闻所未闻。

五、"It"强调句型的译法

英语中常用的强调句型是 It is (was)＋被强调的部分＋that (who, whom, which 等)……,其中 it 本身并无具体含义,在汉译时可以省略,整句通常译为对应的"正是……""就是……"等汉语强调句式,或者将"It is (was)"和"that (who, whom, which 等)"省去不译,将句子还原为陈述句翻译。例如:

1. It was his experience flying the LLTV (Lunar Landing Training Vehicle) that gave him the confidence to take over from the automatic system.

正是驾驶登月训练机的经验让他有信心接管自动系统。

2. It was against this backdrop that Dryden began its innovative lifting body

research in an effort that exemplified the independent, innovative, pragmatic, and pioneer mindset of Dryden's technicians and engineers.

正是在这种背景下,德莱顿开始了其创新的升力体研究。这项工作充分体现了德莱顿技术人员和工程师的独立、创新、务实和开拓精神。

Dryden 指 Dryden Flight Research Center(DFRC)德莱顿飞行研究中心,是美国国家航空航天局的重要研究中心之一。

3. It was, for example, through the unexpectedly poor performance of their 1901 glider at Kitty Hawk that the Wright brothers discovered that some of the basic, accepted tenets of aeronautics of the time were seriously flawed.

例如,正是由于莱特兄弟于 1901 年在小鹰镇的滑翔机试飞出乎意料地表现不佳,让他们发现当时一些公认的基本航空原理存在严重缺陷。

4. It is in the definition of our goals that decision-making for human spaceflight should begin.

载人航天的决策应该从确定目标着手。

课后练习

请翻译下列句子,注意其中的无人称句。

1. It is unclear as to whether SPO (Single-Pilot Operations) is feasible.

2. In many ways, it can be instructive to think of this in terms of an iceberg.

3. It was for this reason that the Dryden Flight Research Center was established in the first place.

4. He stated that it is short-sighted to ignore activity outside of the immediate hub area.

5. It can truly be stated that the F-16XL prototype configurations were in many significant ways the direct result of the joint research effort with NASA.

6. It was important to note that the highly automated function allocations performed very well on some metrics and not so well on others.

7. It was considered unlikely this type of tag would be detected by a hydrophone as the radio frequencies used are very high and quickly attenuated by water immersion.

8. For example, it is reasonable to believe that automation might be more widely accepted as a back-up for the pilot under these circumstances as opposed to the case of pilot incapacitation.

9. It might become an important requirement for the pilot to communicate to ATC that the aircraft is operated by a multi-person crew or a single pilot.

10. It is during this second phase that a technical concept achieves a state of readiness, validation and credibility such that private industry and financing can assume the attendant risks.

11. Engineers from the Johnson Space Center and Dryden agreed that it would be helpful to explore the actual limits and failure modes of the shuttle tires and wheels in real-life conditions, if a suitable test aircraft could be found.

12. Because it was so difficult to protect the tires and gear in the extreme temperatures and environments experienced by the shuttle, the orbiters were equipped with only four small wheels, two on each main gear.

13. But as the nation's focus shifted to space, as transport aircraft stabilized in both configuration and in speed, and as the military's output of new, cutting-edge fighter aircraft decreased, it was harder for outsiders to see the value in a place like Dryden.

14. It had become clear that, while the U. S. might have been the first country to achieve powered, controllable flight of an airplane, government-funded research efforts in Germany, France, and England had allowed European aircraft manufacturers to far outdistance American capabilities.

15. But it was through that focused effort that all the parties involved learned about deadly hazards in the transonic and supersonic worlds, as well as about design flaws and fixes that saved the lives of countless pilots who followed in the footsteps of those first, brave explorers.

第4节　非限定性动词的翻译

为了表述简洁,行文紧凑,科技英语多使用非谓语动词结构代替各类从句。非谓语动词指在英语中不用作句子谓语,而是充当其他语法功能的动词。从英语语法上看,谓语动词的形式要和主语的人称和数保持一致,而非谓语动词则不受此限制,因此又称非限定性动词。

非限定性动词主要包括三大类:动词不定式、动名词和分词(包括现在分词和过去分词)。非限定性动词同时具有动词和非动词特性。动词特性体现在有语态和时态的变化,并且非限定性及物动词可以带宾语和状语;非动词特性指的是非限定性动词不能单独作谓语,可以像名词或形容词等词类一样充当其他句子成分,如主语、表语、宾语、定语和状语等。因此,在翻译非限定性动词结构时,必须要认清句子结构,根据非限定性动词在原句中所充当的成分来具体分析。顺译法、转译法、增译法和分译法是经常使用的四种翻译策略。

一、顺译法

顺译法指的是基本按照原句的语序,将非限制性动词译为汉语中的动词结构。当非限定性动词在句中作主语、表语、宾语、宾语补足语或伴随状语成分时,通常使用顺译法译为汉语中的动宾结构。例如:

1. Finding these relics can yield clues to how our galaxy was built.

发现这些遗迹为了解银河系如何形成提供了线索。

2. Analyzing meteor elemental compositions is crucial to our understanding of how planets like our own develop.

分析流星的元素组成对了解像地球这样的行星如何发展至关重要。

3. Solar radiation is the Earth's primary source of energy, affecting the planet's surface structure and atmospheric conditions.

太阳辐射是地球的首要能源,影响地球的表面结构和大气条件。

4. If desired, ground controllers could command a single camera to remain powered on and for no auto-cycle to take place.

如果需要,地面管制员可以指挥一台相机保持开机状态,避免开启自动循环。

5. With the Hubble Space Telescope, we have begun to understand galaxies as time capsules that chronicle how the universe evolved, from the birth of stars within them to the buildup of immense galaxy clusters.

有了哈勃太空望远镜,我们开始将星系理解为时间胶囊,它们记录了宇宙从星系中的恒星诞生到巨型星系团形成的演化过程。

6. The purpose of the LWF (Lightweight Fighter) program was to determine the feasibility of developing a small, lightweight, low-cost fighter; to establish what such an aircraft can do; and to evaluate its possible operational utility.

轻型战斗机计划的目的是判定研发小型、轻型、低成本战斗机的可行性,确定其用途,以及评估其可能的作战能力。

以上六例中,非限定性动词分别在原句中作主语、伴随状语、宾语补足语、宾语和表语,均可使用顺译法按照原句的顺序将非限定性动词译为汉语中的动宾结构。

二、转译法

由于非限定性动词的非动词特性,在汉译时,可将其译为相应的名词或形容词。当英文原句中的非限定性动词是一个单独的词语,且在汉语中有对应的名词时,可译为名词。非限定性动词在原句中作定语有两种情况,单个的非限定性动词作定语通常为前置定语,非限定性动词短语作定语通常为后置定语,汉译时,一般都处理为前置定语放在被修饰的名词前,并常译为"……的"字结构。当汉语中有相对应的固定表达时,"的"字可省略。例如:

1. At night, with the light of the Sun <u>removed</u>, nature's brilliant glow from Earth's surface becomes visible to the naked eye from space.

夜晚,随着太阳光的<u>消失</u>,地球表面散发出自然的耀眼光辉,在太空中肉眼可见。

2. Glint Mode enhances the instrument's ability <u>to acquire highly accurate measurements</u>, particularly over the ocean.

闪光模式增强了这类仪器<u>获取高精度测量数据</u>的能力,尤其是在海洋上。

3. Astronomers suspected it might be a failed, starless galaxy, or gas <u>falling into the Milky Way from intergalactic space</u>.

天文学家怀疑它可能是一个未能形成的、没有恒星的星系,或者是<u>从星际空间落入银河系</u>的气体。

4. HICO (Hyperspectral Imager for the Coastal Ocean) was the first spaceborne imaging spectrometer <u>designed to sample the coastal ocean</u>.

沿海高光谱成像仪是第一台<u>为沿海海洋采样专门设计</u>的星载成像光谱仪。

5. This is greater than the range of many <u>mapping sonar systems</u> (for example side scan sonar) which may be used to search for an aircraft debris field.

其搜索范围大于许多可用于飞机残骸区搜索的<u>测绘声呐系统</u>(例如侧扫声呐)。

例 1 中"removed"对应汉语中名词"消失",将非限定性动词转译为名词;例 2、例 3 和例 4 分别是不定式、现在分词和过去分词短语作后置定语,译为汉语中的前置定语"……的"结构;例 5"mapping sonar systems"对应汉语中固定表达"测绘声呐系统","的"字省略。

三、增译法

增译法指的是为了完整表达原句意思,按照汉语表达习惯,增译必要的词或句来帮助读者理解。非限定性动词在句中作状语(包括非限定性动词前加上逻辑主语构成独立主格结构的情况)时,可以表示时间、条件、原因、让步、结果、目的、方式和伴随状况,汉译时,可译为相对应的状语从句。除了伴随状语可采用顺译法直接译为汉语中动宾结构外,其他情况应增译那些表示不同类别状语从句的词,例如"当(……时候)""(在……)后""如果""因为……所以""虽然……但是""即使……也""从而""为了""通过"等。例如:

1. This payload's hyperspectral sensor collects hundreds of narrow spectral bands, <u>resulting in</u> extremely high spectral resolution.

飞船携载的高光谱传感器可收集数百个窄谱带,<u>从而</u>获得极高的光谱分辨率。

2. <u>To accomplish</u> both these goals, engineers at NASA's Johnson Space Center (formerly the Manned Spacecraft Center) designed a three-phase test program.

<u>为了</u>实现这两个目标,美国国家航空航天局约翰逊航天中心(前身为载人航天中心)的工程师制定了一个三阶段测试方案。

3. <u>Recognizing</u> that less than 2％ of all investigations could be evaluated，all international partners were engaged in determining which were selected for review.

　　<u>因为认识到</u>可以进行评估的调查研究不到总数的 2％，所以所有的国际合作伙伴都参与确定哪些能被选为审查对象。

4. <u>Conducting</u> a "cosmic archaeological dig" at the center of our Milky Way galaxy，astronomers used Hubble to uncover a population of ancient white dwarfs— smoldering remnants of once-vibrant stars that inhabited the core.

　　<u>在银河系的中心进行</u>"宇宙考古挖掘"时，天文学家利用哈勃望远镜发现了一个古老的白矮星族——它们是一度活跃于星系核心、尚未熄灭的恒星余烬。

5. <u>Measuring</u> chemicals，such as chlorine monoxide radical（ClO），hydroxyl radical（HO2），hydrochloric acid（HCl）and Hypochlorous acid（HOCl）in the middle atmospheric，SMILES（Superconducting Submillimeter-Wave Limb-Emission Sounder）gave the first global observations of the diurnal variations of hypochlorous acid in the upper atmosphere.

　　<u>通过测量</u>中层大气中的一氧化氯自由基（ClO）、羟基自由基（HO2）、盐酸（HCl）和次氯酸（HOCl）等化学物质，超导亚毫米波肢发射探测器首次对上层大气中次氯酸的日变化进行了全球观测。

　　以上五例中，划线部分的非限定性动词在原句中分别作结果状语、目的状语、原因状语、时间状语和方式状语，在汉译时都使用了增译法；斜体部分的非限定性动词在原句中作宾语和宾语补足语，使用顺译法译为汉语中的动宾结构。

四、分译法

　　有时非限定性动词，虽然不在原句中作状语成分，但与上下文有暗含的逻辑关系，也可根据具体情况，把非限定性动词处理成单独的小分句，将其翻译为表原因、条件、目的等的状语从句。例如：

　　1. <u>Being small and light</u> makes transistors more advantageous to vacuum tubes.

　　<u>由于尺寸小，重量轻</u>，故晶体管比真空管更具优势。

　　2. Automatic machines，<u>having many advantages</u>，can only do the jobs they have been told to do.

　　自动化机器虽然有很多优点，但只能按指令运转。

　　例 1 非限定性动词结构"Being small and light"作主语，与后面的句子构成因果关系，所以汉译为原因状语从句，原句的宾语转译为主语；例 2 非限制性动词结构"having many advantages"作"automatic machines"的后置定语，但它表达的意思与后面的句子形成转折关系，因此译为让步状语从句。

课后练习

请翻译下列句子,注意其中的非限定性动词。

1. Measuring the atmospheric ozone and its chemical composition is crucial for understanding the ozone layer depletion.

2. Applying this analogy illuminates a couple of pitfalls to avoid when evaluating the economic value of ISS（International Space Station）research endeavors.

3. There are more than 30 windows with varied optical properties within the ISS（International Space Station）, providing many viewing opportunities for researchers.

4. Although the purpose of the ISERV（ISS SERVIR Environmental Research and Visualization System）was to improve automatic image capturing and data transfer, the images taken in the experiment also aided environmental scientists, disaster responders and other Earth-based users.

5. Remotely sensed data acquired by orbital sensor systems have become vital tools to identify the extent of damage from a natural disaster.

6. At the Great Debate, held before a public audience in Washington, D. C., astronomers Harlow Shapley and Heber Curtis argue whether spiral nebulae are part of our galaxy or outside of it.

7. The results from ASIM（Atmosphere-Space Interactions Monitor）can improve current atmospheric models, including predictions related to climatology, and may improve understanding of the physics of these events and how they relate to lightning.

8. During the surface search, a number of floating objects in the drifted search areas were reported by aircraft, including wooden pallets and fishing equipment.

9. Astronomers have used Hubble's sharp vision to probe the limits of the visible universe, uncovering never-before-seen objects that existed not long after the birth of the universe in the Big Bang.

10. Beginning in 2001, Amgen utilized the microgravity environment during several space shuttle missions to the ISS（International Space Station）to test three drugs.

11. In 1923, Edwin Hubble used the world's largest telescope, perched atop Mount Wilson in California, to measure the distance to the Great Andromeda Nebula, proving that it was much too far away to belong to our galaxy.

12. The ozone layer is the Earth's spacesuit，protecting the ecosystem and the human population by absorbing the Sun's most harmful ultraviolet rays.

13. The purpose of the Astromaterials Division is to combine advancements in science and technology to push human space exploration forward，to apply planetary research，and to develop mitigation methods to establish successful space travel.

14. The ash，made up of tiny pieces of glass and rock，is abrasive to engine turbine blades and can melt on the blades and other engine parts，causing damage and even engine stalls，with consequent danger to the plane's integrity and passenger safety.

第5节　从句的翻译

英语长句可能由一个主句和一个或多个从句构成，使用从句是英语句式扩展的重要手段之一。英语中的从句根据其在句子中的功能不同，可分为三大类：名词性从句、形容词性从句和副词性从句。其中，名词性从句包括主语从句、表语从句、宾语从句和同位语从句四种，功能相当于名词或名词短语，在句中分别充当主语、表语、宾语和同位语；形容词性从句即定语从句，功能相当于形容词，作定语修饰主句的某个名词性成分（先行词），分为限制性定语从句和非限制性定语从句；副词性从句即状语从句，功能相当于副词，在句中充当状语成分，分为时间状语从句、地点状语从句、条件状语从句、原因状语从句、目的状语从句、结果状语从句、方式状语从句、让步状语从句和比较状语从句九种。

为了清晰完整地阐述复杂的概念，准确地描述过程，科技英语中经常出现各类从句。从句的使用是科技英语中长难句较多的主要原因之一，理解和掌握各种从句的译法对于更好地将航空航天科技英语译为地道的汉语至关重要。这一节讨论英语中三大从句的汉译策略。

一、名词性从句的翻译

在名词性从句中，主语从句、表语从句和宾语从句的翻译相对来说较为简单，通常按照原句的语序，使用顺译法即可。同位语从句的汉译，也可使用顺译法，将其译为短句，放在主句后；同时，同位语从句表示与之同位的名词的实际内容，对前面的名词加以补充说明或进一步解释，虽然并不等同于定语从句，但表达的意义和定语从句相近，因此还可以使用转译法，将其译为定语放在与之同位的名词前。例如：

1. <u>What happens next</u> depends on the star's mass—a fate usually sealed at birth.

接下来会发生什么取决于恒星的质量——通常在恒星诞生时,就已注定。

2. The best explanation for the observational data is <u>that the center of the disk is carved out by the action of two black holes orbiting each other.</u>

对这组观测数据的最佳解释是,<u>圆盘的中心是由两个相互绕转的黑洞运动形成的</u>。

3. Scientists suspect <u>that they either occur when two white dwarfs collide or when a single white dwarf siphons material from a companion star.</u>

科学家们怀疑<u>它们要么发生在两颗白矮星相撞时,要么发生在一颗白矮星从伴星吸取物质时</u>。

4. The observation provided the most compelling evidence yet <u>that some supernovas originate in double-star systems.</u>

观测结果提供了迄今为止最令人信服的证据,<u>证明一些超新星起源于双星系统</u>。

5. <u>The idea that planets form in dusty disks around stars</u> is hundreds of years old, but astronomers only found observational evidence relatively recently.

<u>行星是在恒星周围的尘埃盘中形成的想法</u>已有数百年历史,但天文学家最近才从观测中发现证据。

例 1 至例 4,划线部分分别是主语从句、表语从句、宾语从句和同位语从句,都使用顺译法按照原句的语序译为汉语;例 5 中同位语从句进一步解释的名词是句子的主语,译为独立的句子放在主句后不合适,因此使用转译法译为前置定语放在与之同位的名词"idea"前。

二、状语从句的翻译

英语中,状语从句的位置比较灵活,可视情况放在句首、句中或句尾;在汉语中,不同的状语在句子中的位置有约定俗成的用法,跟英语中不同状语从句的位置并不一一对应,因此在汉译时,需要根据汉语的表达习惯,选择使用前置法或后置法来处理。

(一) 前置法

汉语一般习惯将时间状语、地点状语、条件状语、原因状语、方式状语和让步状语置于句首,所以在翻译时,不管这几类状语从句是在原句的句首、句中还是句尾,原则上应使用前置法,译在句首。另外,英语中目的状语从句如果译为汉语"为了……""要(使)……"等,也应译在句首。例如:

1. Nearby stars appear to shift slightly <u>if viewed when Earth is at different points in its orbit.</u>

<u>如果在地球运行轨道的不同位置观察附近的恒星</u>,会发现它们似乎发生轻微的移动。

2. Volunteers of this study had lived in this isolation environment for 56 days when the experiment began.

实验开始时，这项研究的志愿者已经在这种隔离的环境中生活了 56 天。

3. But stars aren't forever；just as new ones are bursting into life on a regular basis，some run out of fuel every day.

但恒星并非永恒；正如新的恒星不断横空出世，每天也会有恒星因燃料耗尽而消亡。

4. In addition，the galaxy likely halted its stellar production because it was starved of fresh，outside material needed to make new stars.

另外，因为缺乏新恒星诞生所需的更多外部物质，该星系可能不会出现新的恒星了。

5. And where we once struggled to fly within our own atmosphere，we're now working on craft that can operate in the atmospheres of other planets as well.

正如我们曾奋力实现在地球大气层中飞行一样，现在我们正在研究可以在其他行星的大气层中同样运行的飞行器。

6. The Loads Laboratory designed a special test fixture along with a series of underwing actuators，overhead hydraulic actuators and cables so that loads simulating distributed air pressures could be applied to the wing.

为了将模拟分布空气压力的载荷应用于机翼，载荷实验室设计了一种特殊的测试固定装置，以及一系列翼下制动器、顶置液压制动器和电缆。

7. Although the data have provided support for some theories of dark matter and spurred debate about the existence of new astrophysical phenomena，more time is needed for the physics community to fully digest the data，and for new or modified theories to be tested.

尽管这些数据为一些关于暗物质的理论提供了支撑，并引发了关于新天体物理现象是否存在的争论，但物理学界需要更多时间来充分消化这些数据，并对新理论或修订的理论进行测试。

以上七例中，划线部分分别是条件状语从句、时间状语从句、方式状语从句、原因状语从句、地点状语从句、目的状语从句和让步状语从句，它们在原句中的位置有前有后，在汉译时都按照汉语表达习惯使用前置法译在句首，其中例 6 中目的状语从句如译为"以便……"，则应放置在句尾。

（二）后置法

不管是英语还是汉语，结果状语从句的位置一般都位于主句之后，因此在翻译结果状语从句时，应使用后置法，将从句译在句尾；若英语目的状语从句译为汉语的"以便……或以免……"等时，也可以置于句尾。

1. Supernovas involve conditions <u>so intense that atoms fuse together to form new elements</u>, which are then strewn throughout the galaxy.

超新星剧烈爆炸，致使原子熔合形成新元素，然后散落在整个星系中。

2. The nose of the F-5 had been radically altered，<u>so that it looked more like a pelican's beak instead of a traditional pointy-tipped，supersonic fighter.</u>

F-5 的机头经过彻底改造，看起来更像鹈鹕的喙，而不是传统的尖头超声速战斗机。

3. All UAVs（Unmanned Aerial Vehicles）must have an FTS（flight termination system）<u>so that，in the event of any loss of control，they can be brought to Earth under control and without harming anyone on the ground.</u>

所有非载人航天器都必须安装飞行终止系统，以便在失去控制时，能够按指令返回地球，而不会伤及地面人员。

例 1 和例 2 是结果状语从句，按照汉语表达习惯使用后置法译在句尾；例 3 是目的状语从句，译为"以便……"，因此也使用后置法放在汉语句尾。

三、定语从句的翻译

定语从句在科技英语中使用十分普遍，有时甚至会出现定语从句套定语从句的情况，导致句子长，结构复杂。因此，需要仔细分析句子的语法结构，找出其内在的逻辑关系，灵活地进行翻译。定语从句虽然用作定语，修饰先行词，但在汉译时，不能简单地把所有定语从句都译为"的"字结构，更不能简单地把限制性定语从句都译为前置定语，把非限制性定语从句都译为分句放在主句后。

仅从语法关系去把握定语从句还不够，还要从功能上去把握。定语从句主要有两种功能：限定功能和描述功能。限定功能包括限制和定义两种含义。限制指的是对先行词在属性等方面加以限制，说明其属性；定义指的是把先行词从所属的类别中区分开来，给以定义。描述功能指的是定语从句描述的是一个事件，并不是对先行词进行区分和限定。因此，在翻译定语从句时，需要分析其内在功能，根据具体情况，灵活使用前置法、后置法或转译法三种翻译策略。

（一）前置法

当定语从句涉及限定功能时，不管是限制性定语从句还是非限制性定语从句，一般情况下都应该使用前置法译为前置定语，放在先行词前；但如果定语从句过长，还是要根据具体情况具体分析，译成通顺地道的汉语。例如：

1. In Glint Mode, the instrument tracks near the location <u>where sunlight is directly reflected on the Earth's surface.</u>

在"闪烁模式"下，仪器会追踪到太阳光直接反射到地球表面的位置附近。

2. A passion for flight research，and pride in their work，runs deep in almost

all the people who work at the Dryden Flight Research Center.

几乎所有<u>在德莱顿飞行研究中心工作的</u>人都对飞行研究充满热情,并为自己的工作感到自豪。

3. The preliminary estimates of the cost of Mars missions are far higher than for other scenarios, all in an era <u>when budgets are becoming highly constrained</u>.

在这个<u>预算日益紧张的</u>时代,火星任务的初步预估成本远高于其他的设想方案。

4. Hubble turned up a nearby galaxy <u>that has remained virtually unchanged for the past 10 billion years</u>—providing a glimpse of what galaxies were like in the very early universe.

哈勃望远镜在附近发现了<u>一个在过去100亿年几乎没有变化的</u>星系,这让我们得以一窥宇宙早期时星系的样子。

5. For example, manufacturing the active suction wing panels, <u>which incorporated millions of laser-drilled holes</u>, was considered impractical and economically infeasible by the aircraft industry.

例如,航空业界认为制造<u>包含数百万个激光钻孔的</u>主动吸力翼板不仅不切实际,也无经济可行性。

以上五例中,划线部分的定语从句均是对先行词进行限定,包括例5,原句中的非限制性定语从句是对要制作的"active suction wing panels"进行限制,所以在翻译时,应使用前置法将定语从句译为前置定语"……的"结构放在先行词前。

(二) 后置法

当定语从句用于描述,对事件进行补充说明时,通常使用后置法,即将从句翻译成并列分句,置于主句之后。使用后置法翻译定语从句时,可以在分句中重复原句中的先行词,或增加"它(他、她)""它(他、她)们""这"等代词代替先行词,也可以省略先行词。例如:

1. Some of that gas cools and precipitates into cold clumps <u>that fall back toward the galaxy's center like raindrops</u>.

其中一些气体冷却、凝聚成冷团,<u>像雨滴一样落回星系中心</u>。

2. The Lunar Base strategy would begin with the building of a base, probably at the lunar south pole, <u>where the Sun is visible much of the time</u>.

月球基地战略将从建造基地开始,基地可能建在月球南极,<u>那里大部分时间都能看到太阳</u>。

3. The finding suggests that quasars might commonly include two central supermassive black holes <u>that fall into orbit about one another after two galaxies merge</u>.

这一发现表明,类星体通常可能包含两个中心超大质量黑洞,在两个星系合并后它们会落入彼此的轨道。

4. Galaxies are cosmic islands of stars, planets, nebulas, gas, dust, and dark matter that are separated from one another in space but collectively help to tell the story of the universe.

星系是由恒星、行星、星云、气体、尘埃和暗物质组成的宇宙岛屿，它们虽散落在太空中，但共同见证着宇宙的发展历程。

5. Astronomers recruited the help of 30000"citizen scientist" volunteers, who sifted through the thousands of Hubble images to search for these star clusters, which range in age from 4 million to 24 million years old.

天文学家招募了3万名"民间科学家"志愿者，他们对哈勃望远镜拍摄的数千张照片进行了筛选，以从中找出年龄介于400万和2400万年之间的星团。

以上五例中，划线部分的不管是限制性定语从句，还是非限制性定语从句，都是在对先行词的性质或者要（可以）做的事件进行描述。这种情况下，使用后置法将定语从句译为并列分句，并根据句意省略或重复先行词，或使用相应的代词。另外，例5中包含了两个非限制性定语从句，斜体部分是对要寻找的"star clusters"的限制，因此要使用前置法译为前置定语。

（三）转译法

有些从句从语法结构上看是定语从句，从语义上来看却有着状语从句的功能。此时，可以根据原句中的逻辑关系，将其转译为原因状语从句、条件状语从句、结果状语从句等。例如：

1. An electric current begins to flow through a coil, which is connected across a charged condenser.

如果线圈连接在带电的电容器上，电流就会开始在线圈中流动。

2. Stars are so bright that thcy typically render surrounding debris invisible, which hides clues about young star systems.

因为恒星非常明亮，光线反差之下，通常无法观测到其周围的碎片，关于新生恒星系统的线索也就无法得知了。

3. RAIDS (Remote Atmosphere and Ionospheric Detection Systems) also studied the ionosphere, which has a strong influence on radio, radar, and satellite navigation signals.

大气层及电离层遥距探测系统（RAIDS）还研究了电离层，因为电离层对无线电、雷达和卫星导航信号都有很大的影响。

4. The arrival of the WORF (Window Observational Research Facility) has allowed astronauts to permanently remove a protective, non-optical "scratch pane" on the window, which had often blurred images.

有了视窗观测研究设施,航天员便可永久移除窗户上的一块起保护作用的非光学"刮擦板",<u>因为它经常会使图像模糊</u>。

例 1 表达的意思是只有连接到带电的电容器上的线圈才会有电流通过,隐藏逻辑关系为满足条件才有相应的结果,因此译为条件状语从句;例 2 中恒星明亮导致周围的碎片不可见,进一步导致新生恒星系统被隐藏,隐藏逻辑关系为因果关系,因此译为结果状语从句;例 3 中研究电离层的原因正是因为它会给无线电、雷达和卫星导航信号带来很大影响;例 4 中刮擦板经常会模糊图像,而这正是宇航员移除它的原因,两个例句的隐藏逻辑关系均为因果关系,因此都译为原因状语从句。

课后练习

请翻译下列句子,注意其中的从句部分。

1. In 1915，Albert Einstein proposed the revolutionary idea that massive objects warp space.

2. It may seem as though the starry sky is a peaceful backdrop full of bright, unchanging beacons.

3. This cloud is supplying fresh material to our galaxy, which feeds new star formation.

4. Hubble, like all telescopes, is a time machine that astronomers can use to see into the past.

5. Transformers cannot operate by direct current, which would burn out the wires in the transformers.

6. A solid fuel, like coal or wood, can only burn at the surface, where it comes into contact with the air.

7. What began as a temporary flight research station became a permanent flight research Center as a result.

8. Another explanation is that the neutron star may emit an energetic wind that slams into gas in interstellar space.

9. Astronomers call it the Leading Arm because it extends out ahead of the Magellanic Clouds in the direction of their motion.

10. Even though the galaxy is 2. 5 million light-years away, Hubble's view resolves more than 100 million individual stars in the galaxy's disk.

11. The analysis indicated that the aircraft's rate of fuel burn was higher during this segment of the flight than first thought and therefore its maximum range was decreased.

12. Initially, management supported telework in concept but was skeptical at

the idea that most of the center's work could be accomplished remotely.

13. Hubble has advanced our understanding of circumstellar environments by using instruments to block starlight so that dimmer features and objects near a star can be seen.

14. The Committee thus identified five questions that could form the basis of a plan for U. S. human spaceflight.

15. The gravitational lens was so strong that Icarus briefly skyrocketed to appear 2,000 times its true brightness, before returning to its original luster.

16. Each "eye" is the bright core of a galaxy, and all of the blue regions are areas where young stars have burst into life as a result of the collision.

17. This distorted galaxy, named NGC 3256, is the product of a collision that occurred roughly 500 million years ago between two spiral galaxies.

18. Although astronomers have located other, brighter galaxies that are slightly farther away, this object represents a smaller, fainter class of newly forming galaxies that must have been common in the early universe but have largely evaded detection.

19. The sonobuoys are equipped with a hydrophone, which is streamed to a depth of 300 m below the buoy and a radio transmitter which allows acoustic data to be relayed back to the aircraft for processing when it is within range.

20. But innovative new vehicles and systems for accessing the atmospheres and surfaces of other planetary bodies were almost certainly going to play an important role in NASA (National Aeronautics and Space Administration)'s future, if the nation wanted to continue its efforts to explore beyond the confines of planet Earth.

第 6 节　长难句的翻译

为了清晰完整地阐述复杂的概念,准确地描述过程,英语科技文本中经常会出现长句。这些句子往往结构复杂,修饰成分多,包含名词化结构、非限制性动词和各种从句,各个成分环环相扣,逻辑紧密,盘根错节。英语长句的翻译难点有时并不在于句子过长,而是句子的各个构成成分之间关系过于复杂,一旦梳理不清或理解错误,就会致使翻译不当。此外,英语和汉语在句法结构上有很大的不同,也给英语长句的翻译增加了难度。

英语重形合,句子成分组织严密,主从句脉络清晰,以主谓结构为主干,其他的修饰成分借助连接词附在句子主干上,构成树形结构,还有学者将英语句子结构比喻成"葡萄型",葡萄主干短,附结的果实多。汉语重意合,句子结构较为简短松散,没有一定的主谓框架,没有谓语动词和非谓语动词的区分,很少有叠床架屋的结构,句子各成分像竹子一样一节一节拼接起来,有学者将其称为"竹竿型"结构。在翻译科技英语中的长句时,要注意这些英汉句子结构上的差异,具体处理方法可以遵循以下的"四步法":

一、切分句子。通读全句,分析句子结构,找出主干,并将原句分割为不同的部分作为翻译单位。

二、初步翻译。理清各部分之间的逻辑关系,运用在第 2 章及本章前几节讨论的关于词、名词化结构、非限制性动词及从句等的翻译策略将各部分译为汉语。

三、调整语序。涉及语序的翻译策略主要有顺译法、倒译法和将两种方法结合起来的综合法,选择哪种译法主要取决于英汉思维的差异。在第 5 节从句的翻译中讨论过,汉语习惯将时间、地点、条件、原因、方式和让步状语置于句首。除此之外,汉语的思维遵循时序原则,习惯按照事情发生的先后顺序进行描述;同时,汉语中重点信息往往后置,次要的信息放在句首,而英语恰恰相反。因此,将分割好的部分译为汉语后,需要按照原句逻辑关系和汉语表达习惯,按需调整它们的顺序,并使用恰当的连接词重新组合成句。

四、检查译文。组织好译文后,对照原文检查是否有漏译错译,逻辑不通或表意不明的地方。

下面具体阐述"四步法"在航空航天科技英语长难句翻译中的应用。

1. It (Hubble) has discovered that galaxies evolve from smaller structures, found
<u>①</u> <u>②</u>

that supermassive black holes are common at the centers of galaxies, verified that
<u>②</u> <u>③</u>

the universe's expansion is accelerating, probed the birthplaces of stars inside colorful
<u>③</u> <u>④</u>

nebulas, analyzed the atmospheres of extrasolar planets, and supported interplanetary
<u>④</u> <u>⑤</u> <u>⑥</u>

missions.
<u>⑥</u>

这个句子虽然属于长句,但结构相对较简单,主要由六个动词并列结构构成,因此可直接将原句分解为以下六个部分进行初步翻译:

① It (Hubble) has discovered that galaxies evolve from smaller structures
哈勃望远镜发现了星系是从较小的结构演化而来的

② found that supermassive black holes are common at the centers of galaxies
发现了超大质量黑洞在星系中心比较常见

③ verified that the universe's expansion is accelerating
证实了宇宙正在加速膨胀

④ probed the birthplaces of stars inside colorful nebulas

探测了彩色星云内恒星的诞生地点

⑤ analyzed the atmospheres of extrasolar planets

分析了太阳系外行星的大气层

⑥ and supported interplanetary missions

并且对行星际任务提供了大力支持

这六小部分的结构都比较简单,①②③的并列谓语后都是宾语从句,④⑤⑥是简单的动宾结构,使用顺译法将各部分译成相应的汉语即可;同时,原句是六个并列结构依次排开,逻辑上同等重要,因此在汉译时没必要调整语序,只需合并后增添必要的关联词,译文如下:

哈勃望远镜不但发现了星系是从较小的结构演化而来的,超大质量黑洞在星系中心比较常见,还证实了宇宙正在加速膨胀,也探测了彩色星云内恒星的诞生地点,同时分析了太阳系外行星的大气层,并且对行星际任务也提供了大力支持。

2. Additionally, such factors as continuing growth in air traffic volume, the
　　　　　　　　①　　　　　　　②　　　　　　　　　　　　　　③

vital role of air transportation on the global economy, and concerns about the overall
　　　　　　　　　　　　　③　　　　　　　　　　　　　　　④

environmental impacts of aviation added focus to the National Aeronautics Research
　　　　　　　④　　　　　　　　　　　　　　　　　　　　①

and Development Policy that was established by President George W. Bush in 2006.
　　　①　　　　　　　　　　　　　⑤

该句是一个主从复合句,主句主语"such factors"后有三个名词化结构作修饰说明,主句间接宾语"the National Aeronautics Research and Development Policy"后跟定语从句进行限定。通读全句后,可将原句分解为下面五个部分并进行初步翻译:

① Additionally, such factors … added focus to the National Aeronautics Research and Development policy…

此外,这些因素使国家航空研究与发展政策更加受到人们关注

② continuing growth in air traffic volume

航空交通量的持续增长

③ the vital role of air transportation on the global economy

航空运输在全球经济中的重要作用

④ and concerns about the overall environmental impacts of aviation

以及关于航空业对环境造成影响的担忧

⑤ … that was established by President George W. Bush in 2006

乔治·沃克·布什总统于 2006 年制定

分割的五个部分中,①是主句主干部分,②③④是对主句主语的进一步解释说

明,汉译时使用直译法译为名词词组放置在主语前;⑤作为主句间接宾语的定语从句,具有限定功能,应译为前置定语,并且该从句中的被动语态有明显的施事者,将施事者作主语转换成主动语态。依照上述分析,将五部分重新排序为①②③④①⑤①,组织语言合并后译文如下:

此外,由于航空交通量的持续增长、航空运输在全球经济中的重要作用以及关于航空业对环境造成影响的担忧等因素,使得乔治·沃克·布什总统于 2006 年制定的国家航空研究与发展政策更加受到人们关注。

3. The changing blue hue that we have learned marks winter at Saturn is likely

　　　①　　　　　　　②　　　　　　　①

due to reduction of ultraviolet sunlight and the haze it produces, making the atmosphere

　　　　①　　　　　　　　　　　③

clearer and increasing the opportunity for Rayleigh scattering (scattering by molecules

　③　　　　　　④

and smaller particles) and methane absorption —both processes make the atmosphere blue .

　　　　④　　　　　　　⑤

这是一个主从复合句,主句主语"the changing blue hue"后跟定语从句,对主语进行详细描述,主句后两个并列的非限制性动词结构做结果状语,破折号后的句子对非限制性动词进行补充说明。通读全句后,可将原句分解为五个部分并进行初步翻译:

① The changing blue hue…is likely due to reduction of ultraviolet sunlight and the haze it produces

蓝色色调的变化可能是由于太阳光中紫外线及其产生的雾霾减少

② …that we have learned marks winter at Saturn

我们已经知晓这标志着土星冬季的到来

③ making the atmosphere clearer

使大气更加干净

④ and increasing the opportunity for Rayleigh scattering (scattering by molecules and smaller particles) and methane absorption

并且给瑞利散射(分子和较小粒子散射)和甲烷的吸收创造了有利条件

⑤ both processes make the atmosphere blue

这两种现象都会使大气变蓝

其中,①是主句,修饰主句主语的定语从句②起的是描述作用,应采用后置法译在先行词后,而先行词"the changing blue hue"在主句和从句中都是主语,因此可按照原句语序将定语从句译为小短句放在主句前,再在主句中重复先行词;非限制性动词结构作结果状语不管在英语中还是汉语中都后置,因此③④顺序不用改变;破折号后的⑤起补充说明作用,汉译时同样使用破折号和主句分开即可。依照上述分析,将

五个部分重新排序为②①③④⑤,组织语言合并后译文如下:

我们已经知晓,蓝色色调的变化标志着土星冬季的到来,这一变化可能是由于太阳光中紫外线及其产生的雾霾减少,使得大气更加干净,并且给瑞利散射(分子和较小粒子散射)和甲烷的吸收创造了有利条件——而这两种现象都会使大气变蓝。

4. Computer simulations fed by Hubble observations of Andromeda's motion
　　①　　　　　　　　　　　　　②

show that the dramatic encounter, predicted to begin four billion years from now,
　　　　①　　　　　　　③　　　　　　　　④

will be a head-on impact, flinging our Sun and its planets into a different part of
　　③　　　　　　　　　⑤

the galaxy, and taking about two billion years to eventually settle into the much
　　　　　　⑥

larger, oval-shaped, elliptical galaxy.
　　　⑥

这是一个主从复合句,由主句、一个宾语从句和两个并列的作结果状语的非限制性动词结构构成,主句和从句的主语后分别又有一个非限制性动词结构作后置定语。通读全句后,可将原句分解为下面六个部分并进行初步翻译:

① Computer simulations show that
计算机模拟显示

② fed by Hubble observations of Andromeda's motion
由哈勃望远镜对仙女座运动的观测提供

③ the dramatic encounter will be a head-on impact
这一令人注目的遭遇将是一次正面碰撞

④ predicted to begin four billion years from now
预计 40 亿年后开始

⑤ flinging our Sun and its planets into a different part of the galaxy
将太阳及其行星投掷到银河系的其他地方

⑥ and taking about two billion years to eventually settle into the much larger,
oval-shaped, elliptical galaxy
大约需要 20 亿年最终形成一个形似椭圆、体积更大的椭圆星系

这六部分中,①和③分别是主句和从句的主干部分,非限制性动词结构②和④分别做①和③主语的后置定语,汉译时应使用前置法译为前置定语放在两个主语前;⑤和⑥是非限制性动词结构作结果状语,顺序无须改变,⑥中"about two billion years"可译为时间状语放在句首。依照上述分析,将六个短句重新排序为②①④③⑤⑥,合并后略微调整语言,译文如下:

基于哈勃望远镜对仙女座运动的观测进行的计算机模拟显示,预计 40 亿年后才

会发生的这一引人注目的遭遇将是一次正面碰撞,届时太阳及其行星会被抛到银河系的其他地方,再过大约 20 亿年后,两个星系最终会形成一个形似椭圆、体积更大的椭圆星系。

　　5. After the 747 (Boeing 747)'s wake induced two unplanned snap rolls in the
　　　　　　　　　　　　　　　　①

T-37 (T-37 Air Force jet trainer) and caused it to develop a roll rate of 200 degrees
　　　　　①　　　　　　　　　　　　　　　　　②

per second despite the fact that the trainer trailed the jetliner by more than three miles,
　②　　　　　　　　　　　　③

one research pilot speculated that a safe separation between the two aircraft in a
　　　　④　　　　　　　　　　⑤

landing configuration would have to be three times that distance.
　　　　　　　　　　⑤

　　该句是主从复合句,由"after"引导的时间状语从句和之后的主句以及一个宾语从句构成。其中,时间状语从句中包含两个并列句和作让步状语的介词结构"despite the fact that",这个介词结构中又有一个同位语从句。句子较长,结构复杂,通读全句后,可将原句分解为下面五个部分并进行初步翻译:

　　① After the 747(Boeing 747)'s wake induced two unplanned snap rolls in the T-37 (T-37 Air Force jet trainer)
　　波音 747 的尾流导致 T-37 空军喷气式教练机发生两次突然的意外翻滚后
　　② and caused it to develop a roll rate of 200 degrees per second
　　并且导致其翻滚速度达到每秒 200 度
　　③ despite the fact that the trainer trailed the jetliner by more than three miles
　　尽管教练机跟在喷气式飞机后超过 3 英里
　　④ one research pilot speculated that
　　一名研究飞行员推测
　　⑤ a safe separation between the two aircraft in a landing configuration would have to be three times that distance
　　在着陆时,两架飞机之间的安全距离必须是该距离的 3 倍

　　本句中,"after"引导的虽然是时间状语从句,但是它和主句的隐含逻辑其实是因果关系,主句是基于从句描述的事件推测而来的,所以在翻译时,可以转为汉语中的结果状语从句;在①②③组成的时间状语从句中,③是让步状语,应译在句首;宾语从句⑤中,"in a landing configuration"是时间状语,应译在句首。依照上述分析,将五个部分重新排序为③①②④⑤,合并后调整语言,译文如下:

　　尽管 T-37 空军喷气式教练机和波音 747 之间的距离超过 3 英里(4.8 千米),波音 747 的尾流还是导致 T-37 发生两次突然的意外翻滚,并且致其翻滚速度达到每秒

200 度,因此一名研究飞行员推测,在着陆时,两架飞机之间的安全距离必须是该距离的 3 倍。

课后练习

请翻译下列长句,注意长句的翻译方法和策略。

1. The easiest way to solve this was to invent the engine and the the crankshaft/propeller shaft at the top of the engine, quite different to automotive engine design where the crankshaft is always at the bottom.

2. This research was supported by"Lunar Palace 365", which was a 370-day, multi-crew, closed experiment carried out in a platform with ground-based experimental bioregenerative life support system (BLSS), named Lunar Palace.

3. This was surprising, because that much sulfur can only be supplied by previous generations of stars, which would not have existed in a cloud of pristine hydrogen from outside our galaxy or in the remnant of a failed galaxy lacking stars.

4. Observations made with Hubble and other telescopes suggest that spiral galaxies such as the Milky Way and Andromeda grew as dwarf galaxies merged, and that they continue to grow larger even now by pulling in and absorbing these smaller galactic satellites.

5. Researchers would also attempt to determine the potential benefits of various advanced aircraft configurations beyond the conventional tube-and-wing design that has been standard since the earliest days of commercial air transportation.

6. NASA researchers believe that in the second quarter of this century, commercial airline companies could save as much as $250 billion thanks to so-called "green" aviation technology pioneered by the Agency and industry partners under NASA's Environmentally Responsible Aviation (ERA) project.

7. All of the"configuration" research aircraft exhibited unique quirks and problems, but all made important contributions to the aerospace industry: the X-4, for example, oscillated severely about all three axes as it approached Mach 0. 9, convincing researchers that a semi-tailless configuration was altogether unsuitable for transonic or supersonic aircraft.

8. If, for lack of funding or support, good people at Dryden leave, or the center's testbed aircraft are sent to the bone yard, the capacity of Dryden to conduct a successful X-plane research, support industry's efforts to advance technology, help NASA develop new spacecraft designs, or even support efforts to make

existing aircraft or spacecraft safer, will be compromised.

9. By independently developing an experimental prototype that would inexpensively validate the concept of transonic/supersonic cruise and maneuverability along with improved air-to-ground capabilities, GD hoped to interest the Air Force in supporting development and production of what was essentially a new aircraft, but one that shared much in common with the basic F-16.

10. Its small size, pragmatic focus, and its informal, flexible, innovative style of management and operation have helped encourage the development of a staff infused with both technical passion and technical agility-traits that have allowed the center to support a wide range of research programs and priorities, as well as adapt to the constantly changing times and priorities of the past 60 years.

第4章　航空航天科技英语翻译中的修辞与逻辑

科技文本是科学与文学的结合体,既有科学客观的内在信息,又呈现出地道自然的文字表达艺术。想要做好科技文本的翻译,必须内外兼顾,在传达信息时要做到客观准确,条理清晰,逻辑严密;在文字表达上力求生动形象地反映客观事实,增添文学艺术美感。本章将从科技翻译中的艺术性、逻辑性两方面入手,探讨如何在翻译实践中,用生动传神的文字让高深复杂的信息变得明晰易懂,在"信"的基础上,兼顾"达"与"雅"。

第1节　科技翻译中的修辞

许多人认为,科技文本用以传达理论概念,行文规范严谨、注重客观、表达平易,因而不应掺杂任何艺术性因素和修辞性手法,以免有损客观。相应地,其翻译也应谨遵原文风格,把信息从译出语传送到译入语即可,无须考虑思想感情、美学体验、文采修辞。这种认知未免有些狭隘。无论何种文体,可读性和易懂性都是最基本的要求。而对于多专业术语、深奥概念的科技文本而言,达到这一要求更是难上加难。想要让叙述说理清晰易懂,让读者群体不局限于专业人士,免不了修辞手段的运用。

一、修辞的定义

修辞并不仅指比喻、拟人、排比等狭义的修辞格,而是一个更为广义的概念。张志公曾说,"修辞就是选择"。广义的修辞包括选词、择句、设格、组段和谋篇五个过程,覆盖由词汇到篇章的各个层次。吕叔湘也将修辞定义为,"在各种可供选择的语言手段之间——各个(多个是同义的)词语之间,各种句式之间,各种篇章结构之间,各种风格之间进行选择,选择最合适需要的,用于达到当前特定的目的"。这样看来,修辞在科技文本中的运用是自然而然的。

在《修辞学发凡》中,陈望道将修辞进一步细分为消极修辞和积极修辞。消极修辞是"抽象的、概念的",主要包括用抽象思维方法,运用概念进行合乎逻辑规律的推论,在语言文字的形式方面把要表达的内容意义明确、伦次通顺、词句平匀、安排稳密地表达出来;而积极修辞是"具体的、体验的",指的是用形象的思维方法和艺术的描绘手法,生动形象地表达思想内容。与积极修辞相比,消极修辞常因表现方式依托逻辑表达而不够明显,甚至难以辨认。但在科技文本中,运用最为广泛的却是消极修辞。逻辑层次上的思维缜密,内容表达上的练达晓畅,结构布局上的均衡妥帖,都属消极修辞范畴。以下几个例子就是消极修辞在科技文本中的应用：

1. The horizontal forces an airplane is subjected to during its flight are thrust, which is generated by the engines, and drag, which is caused by friction and differences in air pressure.

飞机在飞行过程中受到推力和阻力两种水平方向的力。推力由发动机产生，阻力由摩擦和气压差产生。

例1中的原句是典型的"树式"复合长句，在中心词 thrust 和 drag 后分别紧跟一个 which 从句对新概念进行解释。但汉语中通常先整体描述并列成分，再分别解释，以达到整齐对称的效果，多用短小单句。因此，在翻译时将宾语 thrust 和 drag 和各自的非限定性定语从句拆分开来，将从句处理为两个独立小句放在主要信息后进行补充。这种对句子结构进行的"化整为零"就是运用消极修辞的体现。

2. Rectification of this fault was achieved by insertion of a wedge.

嵌入一个楔子便纠正了这一误差。

例2中出现了科技文本中常见的被动句和名词化结构。句中将动作的主客体顺序倒置，利用"was achieved by"这一被动结构凸显文本的客观性，但汉语中带"被"字的结构不常用，因此，在翻译时将被动结构还原，按主宾顺序顺次翻译。原句中的"rectification"和"insertion"都是由动词转化而来，由于中文更习惯使用"动宾短语"，在翻译时便将名词化结构还原，还作为句子中的动词成分。这样一个短句的汉译中，运用了句式转换、化静为动两种消极修辞手法，由此看来，消极修辞在科技文本汉译中的运用可见一斑。

3. Not only do these satellites perform a variety of tasks but they come in numerous shapes, sizes and nationalities.

这些卫星不但执行的任务多种多样，而且形状有别，大小不等，国籍迥异。

英语中常见"多枝共干"式表述，即一个动词连接多个名词，或一个形容词修饰多个名词。例3中画线部分便是由形容词"numerous"修饰了三个并列名词。为了符合汉语富于变化、对偶齐整、朗朗上口等特点，译文对搭配中的多义词进行变换，形成三组不同的配合关系。

消极修辞贯穿于科技文本翻译的遣词、造句、构段、谋篇所有过程，在很大程度上表现为对所译词语和句式的灵活选择和巧妙运用，以确保译文尽可能传达原意、符合表达习惯、贴合内在逻辑。与消极修辞关注意义的表达不同，积极修辞着重的是对外在形式的雕琢，通常由修辞格表现。随着科技英语的通俗化，积极修辞的运用也越来越多。以下我们将从原文理解和译文表达两方面进行细致讨论。

二、科技翻译与修辞

理解和表达是翻译过程中的两个主要阶段，翻译的本质就是正确理解原文和准确再现原文的动态过程。可以说，理解或表达稍有差池，都会影响译文的质量。在科技翻译中,修辞的运用加大了理解和表达的难度。本节将分别对这两个阶段如何处理修辞部分进行讨论。

(一) 修辞与原文理解

在任何文体的翻译中,正确理解原文都是最基本的要求。没有正确的理解,何谈准确的表达。虽然使用修辞的初衷是将信息以更为通俗易懂的方式传达给读者,但有时也会造成理解上的障碍,例如:

Is cloning technology becoming the sword of Damocles to human beings?

如果译者在阅读这句原文的时候没有认识到"the sword of Damocles"其实是运用了比喻的手法,将会看得云里雾里,完全不知此句所云,进而无从下手翻译。即便通过语义猜出了这部分是喻体,不了解这背后的故事也无从判断这一把"剑",究竟是对人类有益的剑还是有害的剑?

达摩克利斯之剑(the sword of Damocles)源自古希腊传说,指临头的危险。上句可以译为:

克隆技术是否日益成为人类头上的一柄达摩克利斯剑,无时无刻不在威胁着人们的安全呢?

译文在将"the sword of Damocles"直译为达摩克利斯剑后,又增译了"无时无刻不在威胁着"进一步解释,表达出了作者以达摩克利斯剑喻指的隐含意义,并通过增补的消极修辞手段,显化这一深层意义,让读者能全面了解其中隐含的深层逻辑。译文既保留了原文的艺术性,又说理清晰,没有因顾及艺术性而有损客观性。

再如:

Tumors fall to a Trojan horse. Altered bugs can carry a lethal enzyme to skin cancer cell.

如果不了解"Trojan horse"的含义,很可能会把这句话误解为"某种马得了肿瘤"。其实"Trojan horse"就是常说的"特洛伊木马",喻指内部破坏者。因为这一典故广为人知,因此直译为:

肿瘤变成了一具"特洛伊木马"。肿瘤组织内发生变化的病菌可能携带一种能够杀死皮肤癌细胞的酵素。

在翻译实践中,要注意句中可能运用的修辞格,如果在原文中遇到了理解障碍,应首先判断作者是否运用了修辞手法;其次,在确定了具体修辞手法后,要勤查资料,全面了解用以修辞的部分(如喻体)的含义,找到其与本体之间的内在逻辑,再选择合适的方式将其清楚明白地译出。

(二) 修辞与译文表达

准确理解文意后,将已经意会的信息忠实地用另一种语言表达出来,在不破坏文体风格的前提下,实现语言的艺术性,增加译文的美感,激发读者的阅读兴趣。

做翻译工作要词典不离手。在翻译实践中，在理解文意阶段需要词典的辅助，在锤炼语言时更要勤查词典。但这一阶段查词典的目的不仅是要找到对应的释义，更多的是通过词典中的释义、用例启发思维，让译者能跳出惯性思维的局限，在选词上发挥主体性进行合理创造，以灵活变通的方式让叙事说理更为生动易懂。例如：

1. On August 6th a <u>flimsy-looking</u> pilotless aircraft, the Zephyr S, came slowly into land at an undisclosed location in Arizona.

句中出现了合成词"flimsy-looking"。查词典后可知，flimsy 这一形容词释义有：①劣质的，不结实的；②（材料）薄而易损坏的；③不足为信的。而这句中明显应取②之义。但如果生搬硬套词条释义，译为"一架薄而易损的无人驾驶飞机"，会十分生硬怪异。重看词典，发现释义②下常用搭配有 a flimsy piece of paper（薄薄的一张纸），受此启发，将"单薄"之意借纸这一意向表达，得出通顺合理的译文：

8月6日，一架薄得像纸的无人驾驶飞机"西风 S"号缓慢降落在美国亚利桑那州的一个未知地点。

例句的原文中并无比喻修辞的运用，但在翻译过程中，为了准确传达"flimsy"一词的意义，且保证译文流畅地道，将这一句灵活处理为比喻修辞，不仅没有扭曲文意，有损客观，还通过"纸"这一喻体凸显了飞机的单薄脆弱，给读者以更深刻直观的阅读感受，可谓精妙。

2. Scientists have all along believed that some extreme physics might take place in the innards of a neutron star, yet <u>there is considerable confusion</u> as to these processes will generate what kind of new matter.

一直以来，科学家们都相信中子星内部会发生某些极端的物理现象，但对于这些现象会产生何种新物质，则<u>众说纷纭，莫衷一是</u>。

"considerable confusion"如进行字对字翻译，为"相当大的困惑"，虽然也算通顺，但显得较为平庸。中文表达一向推崇整齐和谐，句中多采用对偶排比、对立并联。基于这一语言习惯，译者用两个短小精悍的成语构成对偶修辞，既强调了语义，又实现了字音和字节的匀称协调，为译文增色不少。

从以上两例来看，在科技文本的汉译中，运用合适的修辞手段对原文进行灵活表达不失为一种既保证"信"又实现"雅"的好方式。

三、科技文本中的修辞格运用与翻译

本节将对科技文本中较为常用的修辞格进行举例分析。

（一）明喻（Simile）

明喻，即带有 like, as, as if 等明显比喻词的比喻，其通过将新概念和已知概念的相似性加以联系，让新概念变得更为通俗易懂。对于包含明喻的句段，科技英语翻译中通常采用直译。

1. Communications satellites string <u>like beads</u> in geosynchronous orbit，while low-flying satellites crisscross the globe <u>like busy bees</u>，connecting folks too far-flung or too much on the go to be tethered by wires.

通信卫星像珠串一样与地球在同步轨道上运行，而低飞的卫星就像忙碌的蜜蜂往返穿梭于全球，使得因相距遥远或行踪不定而无法使用有线通信器材的人们相互取得联系。

2. Like a solid hit fast ball，the Big Bang is going-going-going.

像一个被猛击一下飞速远去的球，大爆炸正渐行渐远，一去不复返。

例1中将通信卫星比作"珠串"，低飞的卫星比作"忙碌的蜜蜂"，生动形象地凸显出了卫星数量之多；例2中将大爆炸的远去比作"球被击飞"这一动作，给人以运动之感，大爆炸远去的过程跃然纸上。对于这两个由 like 连接的比喻成分，都采用了直译方法。

（二）暗喻（Metaphor）

暗喻与明喻相对，在句中不出现任何比喻词，通过读者自行联想增进理解。许多科技术语都是由暗喻得来，如描述火箭或导弹发射时的 blastoff。对于暗喻的翻译，有时采取直译方法，有时要对喻体进行解释，如何处理要视情况而定。例如：

1. But as the Web grows beyond <u>infancy</u>，many of the world's leading telecommunication and Internet companies are aiming to cut the Web's <u>umbilicalcord</u>.

不过，随着因特网日趋成熟，度过其"幼儿期"，许多全球领先的电信和互联网公司正在想法要将其"脐带"切除。

2. The <u>babel</u> of their plan to invent a perpetual motion machine must terror to the ground.

他们想发明永动机的计划是空想，到头来必然落空。

例1中，"幼儿期"指发展初期，这一深层含义基本人人都能理解，而且由于和后文的"脐带"形成的前后照应，因此在翻译时选用直译方法；而例2中的"babel"巴别塔：巴别（巴比伦的示拿的古城，诺亚的子孙拟在此建一座通天塔，上帝怒其狂妄，乃乱其语言，使建塔人突操不同的语言而四散，塔因此终未建成）[《创世记》11：1～9]是比较陌生的概念，读者很可能不清楚巴别塔意味着什么，因此在翻译时将其背后的深层含义解释出来，省略了巴别塔这一喻体，为了保证可读性牺牲了艺术性。

（三）拟人（Personification）

拟人，即将人类行为赋予无灵事物。因为现代科技中许多机器的最终目标就是模拟人的各种行为，因此许多科技术语都是由拟人这一修辞得来的，如计算机的 memory（记忆器）、机器人的 arm（机械臂）等。在科技文本中使用拟人修辞，可增加所描写事物的活泼感,给严肃平实的科技文体增添一丝趣味。对于拟人部分的翻译,有时选择直译,保留其艺术效果;有时则进行解释,保证意思通顺。我们以出现在一篇新闻报道中的两句话为例:

1. Hubble telescope sees a space "snowman" thousands of light-years away.

哈勃望远镜在数千光年外"看"到了一个太空"雪人"。

2. The Hubble Space Telescope's sharp eyes picked up the object from a distance of 6000 light-years away, and rendered the image in a time exposure since the glow of the gas is very faint.

哈勃望远镜敏锐地锁定了这一6000光年外的物体,但由于这团气体发出的光芒非常微弱,所以只能用定时曝光的方式对这张照片进行渲染。

在例1中,只有有灵生物才能做到的"see"被直译为"看到",既表现出了哈勃望远镜是用于观测的实质,又使其产生了具象效果,拉近了和读者间的距离;但如果在例2中才采取直译的方式将译文处理为"哈勃望远镜敏锐的眼睛锁定了……",会徒增奇怪之感。因为望远镜没有眼睛,也没有和眼睛类似功能的组件,因此在翻译中选择将"eye"这一意象抹去不译。

(四)类比(Analogy)

类比和比喻常常被人混淆,比喻比较的是类似事物间的共同点,而类比比较的则是两种本质上截然不同的事物之间的共同点,通过列举具体形象、人们熟知的概念使抽象难懂的新概念更易接受。类比常出现在科普文章和科技新闻中,如刊登在Space Daily上一篇名为Next Stop:Planet Mars的新闻中,在形容让飞船准确降落在预定位置的艰难时写道:

"It's like playing golf … where you tee off in Paris and the hole is in Tokyo," said navigation team chief Louis D'Amario. "It's the equivalent of locating something the size of a quarter in New York City, seen from Los Angeles."

路易·达马里奥(Louis D'Amario)是导航系统的负责人,在他看来,"这就像打高尔夫时……你在巴黎发球,而球洞在东京;或是在洛杉矶用千里眼在纽约市找一个25美分硬币大小的东西。"

为了让常人理解技术层面的难度,这一负责人连用两次类比修辞:先是将其比作横跨巴黎东京打高尔夫球,既显示了路程的遥远,又让人清晰地体会到想要"一杆进洞"准确降落有多不容易;紧接着和远在3 000千米外找极小的硬币相比,更进一步体现出了对精度的要求。

(五)头韵(Alliteration)

头韵是英语作为字母语言特有的修辞格,是利用发音制造特殊效果的一种修辞方式。简单来说,头韵就是两个单词或两个以上单词的首个字母和发音都相同。头

韵在英文诗歌中的应用较为普遍,在科技新闻标题中也常有体现,例如:

S̲olar S̲ystem S̲earch from S̲pace

从太空探索太阳系

S̲pace S̲huttle's S̲ecret Alter Ego:Crime Fighter

航天飞机不为人知的另一面:犯罪斗士

Coronavirus Delays M̲anned M̲ars M̲ission

新冠疫情推迟了载人火星任务

从上述例子可见,头韵在翻译的过程中基本无法保持。这是由于中英两种语言在构成上具有本质性不同,因而头韵这一修辞格基本无法在汉语中找到对应修辞方式。况且汉语中多习惯押尾韵,而较少注重头韵,因此无须纠结于此,只需把原文中的意思忠实地表达出来即可。

(六) 反复

反复,即相同的词或短语在句中的反复使用。一般来说,反复多见于文学文本,以增强句子气势和感染力。科技文本中,反复的使用多见于产品的使用手册、说明书,如:

The following operational conditions may be appropriate:

a. device under specified load and operating conditions;

b. device under full load (if different from a.);

c. device under no load;

d….

适用的运行条件如下:

a. 装置处于规定负载和运行条件下;

b. 装置处于满负载下(如果情况与 a. 不同);

c. 装置处于无负载下;

d….

除了上述的六种修辞格,低调陈述、排比、省略等修辞格在科技文本中也会有所涉及。在进行修辞格的翻译时,首选汉语中相应的修辞格,但由于英汉两种语言间的差异,很可能无法准确地一一对应,这种情况下为保证语义的忠实畅达,只能选择转换修辞格甚至省译修辞格。因此,具体情况还是需要译者自行判断,灵活处理。

课后练习

将下列英语句子翻译成汉语,注意句子中的修辞使用。

1. NASA's new space telescope has captured its first starlight and even taken a selfie of its giant, gold mirror.

2. As a ferryman can steer his boat orthogonally to the river's current，so the planets could move in and out with only a constant sideway current.

3. He (Kepler) postulated that the sun rotated around its axis，creating a whirling circular river of motive power that pushed the planets around.

4. NASA released the selfie，along with a mosaic of starlight from each of the mirror segments. The 18 points of starlight resemble bright fireflies flitting against a black night sky.

5. "Stars can survive being stretched a small amount，but this star was stretched beyond its breaking point. This unlucky star just wandered into the wrong neighborhood，" said Dr. Komossa.

6. Correct timing is of the utmost importance，and also extreme accuracy, down to a twenty-thousandth part of an inch，in the grinding of certain parts of the fuel injection pump and the valves.

7. Fatigue failure of structural components of an aircraft of fail-safe design is quite acceptable，provided it does not occur often enough to endanger the aircraft, reduce its service life，or reduce its utilization and economy by excessive maintenance.

8. The fans are often fitted with movable shutters to their air intakes which open and close automatically under the control of thermostats to keep the cylinder temperatures as even as possible，admitting more air when the engine is working hard and less when it is idling.

9. Associated with limit loads is the proof factor，selected to ensure that if a limit load is applied to a structure the result will not be detrimental to the functioning of the aircraft，and the ultimate factor，which is intended to provide for the possibility of variations in structural strength.

10. Telstar was followed rapidly by other satellites of increasing sophistication, notably Early Bird 1，the first commercial communication satellite，which was launched in 1965 and is hovering some 22，300 miles above the Atlantic；it regularly carries commercial telephony or television between Europe and America.

第 2 节　科技翻译中的逻辑

翻译是对词汇、语法和修辞等语言问题的处理，也涉及逻辑等一些非语言因素。从一定程度上说，逻辑是翻译过程中最重要的因素，其重要性既体现在原文理解的过程中，更体现在译文的表达上。这一点在科技翻译中尤为突出。在科技翻译中，透彻地理解原文离不开逻辑分析和判断,精准确切、条理分明的表达更得依靠逻辑才能达成。

在讨论科技翻译中的逻辑之前,先了解一下篇章的概念。篇章是比字、句、段更高一级的语言概念,是结构和翻译的最大单位。篇章传递的往往是完整的信息,因此译者在翻译时如能从篇章角度考量,就可以对文本产生一定的整体性认识,避免因"只见树木,不见森林"犯下错误。可以说,翻译的最终目标就是求得译文篇章的意义相符、功能相当。

英国语言学家夸克(Quirk)将篇章定义为"实际使用中的得当连贯的语言片段",表明篇章既是基于语法现象的"语言片段",又要符合"实际使用"时"得当连贯"的要求;博格兰特(Beaugrande)认为,"篇章可以定义为符合衔接、连贯、有目的、可接受、含信息、含情景和互文性这七条篇章性标准的交际行为";韩礼德(Halliday)也指出,"判断一系列句子是否构成了一个篇章,取决于句内与句间的语义连贯关系",而句内与句间的语义连贯关系则是通过逻辑上的联结表现的。因此,译者在对篇章进行整体性分析时,不仅要关注篇章内部的语法结构,更要将篇章外部的逻辑关系纳入考量,把篇章结构置于篇章类型所对应的逻辑视角下进行综合研究。

篇章的形成和逻辑密不可分。篇章的语义关系是以逻辑关系为基础的,并由特定的逻辑关系显现出来。何善芬认为,逻辑在篇章中至关重要:"形成语篇的根本是逻辑,理解语篇的根本也是逻辑,一切语篇无不深藏着思维的逻辑。无论是显性的连接还是隐性的连贯,可以没有语法或词汇衔接,但决不可没有逻辑上的衔接,否则便无完整的语义整体,也就无所谓语篇了。"因此,一篇文章想做到前后呼应、层次分明、通顺易懂,必须基于逻辑进行谋篇布局;相应地,想要透彻理解一篇文章的意义,也必须由捋顺篇章逻辑入手。

一、英汉篇章逻辑比较

受不同的思维习惯影响,英汉谋篇布局的方式也截然不同。英语是语法型语言,有鲜明的词性分类、时态标志、语法结构,注重以形显义,句子结构层级严谨;而汉语则是典型的语义型语言,注重以意役形,句子成分之间靠隐性连贯、逻辑关系和叙述的事理顺序联系在一起,句子结构松散自由。英语重形合,强调外部语言形式上的对应;而汉语重意合,更看重行文意义上的连贯。英语句子突出主语,严格依照"主谓"框架形式展开,句中各成分相互制约,如同参天大树,在主干结构(主谓)的基础上延伸叠加;而汉语构句则突出话题,遵循"话题-说明"的逻辑,构造呈"竹型"结构,一个句子可由好几个短句组成,而各短句之间只有意义上的联系,不需要形式上的显性勾连。

以中英文中对"压差阻力"的定义为例:

英语 Pressure drag is drag caused by increased pressure on the front and decreased pressure on the rear of an object moving through a fluid medium such as air or water or of a stationary object around which the medium passes.

汉语 相对气流流过机翼时,机翼前缘的气流受阻,流速减慢,压力增大;而机翼后缘气流分离,形成涡流区,压力减小。这样,机翼前后产生压力差形成阻力。这个阻力称为压差阻力。

通过以上对比可以看到,英语定义在开头便用四个单词构成的主系表结构直接定义了压差阻力的性质,即压差阻力(pressure drag)是一种空气阻力(drag),形成句子主干,让人初读便对压差阻力是什么有了初步的理解;接着在句子主干后通过后置定语(caused)交代压差阻力形成原因,利用这一语法结构为主干补充信息;在指出压差阻力的作用对象——物体边缘时,也是先写出 object,再用动名词 moving 对具体是什么样的物体进行更为细致的限定。

而汉语定义在叙述时则先以"相对气流流过机翼时"这一时间状语进行语境铺陈,交代压差阻力形成的条件,让人乍一读并不能明确这句话的中心思想;接着在状语限定的条件的基础上分别对机翼前缘、机翼后缘发生的现象进行描述,完全是按照思维逻辑自然推进;前缘和后缘的压力一个增大、一个减小,读者读至此已经可以推断出二者间产生了压力差,这时句中才明确给出"压差阻力"这一概念,明确主旨。

由以上例子可以看出,英语篇章结构通常以"总分"形式呈现,开篇便直截了当点明中心主旨,再通过语法结构从各方面逐级扩展细化。而汉语则很少开门见山,而是先进行背景铺陈,而后按逻辑发展顺序如流水般自然推进,在结尾总结时才点出主旨。再如:

英语 The purpose of this book is to place at the disposal of the design engineer who is facing these challenges, a survey of the experience gained from the many and diverse applications of aerostatic bearings which have already been successfully accomplished

汉语 空气静压轴承已经研制成功,在各个方面的应用中也都取得了很多经验,因此,向承担这些任务的设计工程师进行介绍便是本书的目的。

以上英语长句在开篇便交代了中心主旨,总体描述"本书的目的"是什么,先出主干,再在主句后插入定语从句和分词短语,解释为什么目的会是这样;而汉语则不同,为符合先因后果的思维方式,先对这本书的科技成果背景进行介绍,正因为取得了这样的科技成果,才使得这本书得以完成,具有某种目的,这种表述完全是按照逻辑发展顺序组织的。

二、逻辑在科技翻译中的重要性

翻译是一种包括思维过程和表达过程的高度复杂的脑力活动,一时一刻也离不开逻辑。匈牙利翻译家拉多久尔吉称翻译就是逻辑活动,翻译作品就是逻辑活动的产物。翻译由理解和表达两方面构成,无论是在理解方面还是在表达方面,逻辑都起到非常关键的作用。对原文的理解实际上是语义辨析、语法分析和逻辑分析三个方面的交互过程。英语被称为含糊语言,即外延不确定、内涵无指定的语言。其在本质上是明确的,但在表象上是模糊的;在内容上是确指的,但在形式上是灵活的。因此,在处理英语篇章时,如果仅从表象入手、只分析形式,很容易产生理解偏差。想要准确地理解到作者想表达的表层和深层含义,必须深入语篇内部,依托逻辑分析。

得出精准达意的译文的前提,是要对原文有准确的理解。科技英语往往句式嵌套、层次复杂,如果没能正确将顺各部分间的逻辑关系,很容易造成误解、误译,例如:

Shortly after the unhibited space station reached orbit in May 1973, aerodynamic pressure ripped off a meteotid and heat shield.

原译 在 1973 年 5 月无人空间站到达轨道前不久,空气动力压力扯掉了<u>一个流星体和挡热板</u>。

这一译文中出现的误译连非专业人士都能一眼看出,空气动力压力怎么能扯破流星体?显然译者是将"meteotid"和"heat shield"当成了并列成分。但从语法上看,如果二者真为并列成分,那么在 heat shield 前也应有一不定冠词"a"。因此,这两部分并不是并列的逻辑关系,"a"修饰的也不是"meteotid",而是"shield,meteotid"和"heat"才是并列关系,共同修饰"shield"。

改译 在 1973 年 5 月无人空间站到达轨道前不久,空气动力压力扯掉了<u>一个防流星体和防热的护罩</u>。

在形成了正确的理解后,下一步就是要组织译文。科技英语文本的严谨性和逻辑性要求译者在翻译时,绝不能仅是机械性地逐词翻译、连词成句、垒句成段,而是要从整个语篇逻辑框架出发,在整体的一致性中选择合适的词义、句法、段落排布结构。机械的字对字翻译只会让译文看起来生涩怪异,翻译腔严重,例如:

Fuel density is variable and fuel is sold on a volumetric rather than a weight basis, and for our purposes it will be considered sufficient to use the standard density shown.

原译 <u>燃油密度是可变的,且燃油是按体积而非重量出售的,暂且认为使用所示的标准密度即足以实现我们的目的。</u>

以上译文几乎是字对字的死译,译文信息重复冗余,翻译腔明显,不符合汉语的语言逻辑。通过逻辑分析可知,原文由两个层次的并列句组成,第一个层是逗号前"and"连接的两个短句,是递进的逻辑关系;第二个层是逗号前后两个分句,存在转折关系。为显化这两层逻辑关系,在组织译文时需要进行适当的增译,以准确表达出原文的含义。修改后的译文如下:

改译 燃油密度是可变的,<u>而且</u>是按体积而非重量出售的,<u>但是,我们认为,在飞机设计中</u>,使用所给出的标准密度已经足够了。

三、科技英语的逻辑翻译策略

一般说来,科技英语篇章中的叙事逻辑为:首先开门见山地提出中心论点,而后层层递进逐一展开论证各分论点，作为中心论点的有力支撑，最后通过总结提出推理结论。而汉语科技文本中的叙事则往往与此相反，具有以下特征：①先通过状语成分限定叙事的时空或逻辑条件，进行背景铺陈，再出主题；②先按步骤描写现象过程，让读者对某一概念形成一定认识后，再下定义；③先罗列数据、阐述事实，通过充分的实证自然引出结论；④先交代现象发生的原因，通过因果逻辑自然推出相应结论。例如：

1. Some immediately raise objections to the suggestion that there is a trade-off between swept back wings and stealth features.

后掠翼和隐身功能不可兼得，有些人立刻对这一观点提出异议。

2. He was helped by the fact that the metal tip is very clearly visible on X-rays and so its progress into the tube can be watched and guided very precisely.

在 X 光下，金属顶端清晰可见，因此，他可以借助 X 光精确地观察和引导金属顶端进入导管的全过程。

例 1 中，英语先表达中心思想，即"有些人提出了异议"，再利用同位语从句交代提出异议的对象；而汉语中则先阐述异议产生的先决条件，再自然而然地引出"提出了异议"这一结论，英汉语序完全相反；例 2 中，英语先说结果，即"他得到了帮助"，再利用同位语从句补充产生这一结果的原因；汉语中先交代原因，并且加入了因果逻辑连词"因此"，再对结果进行总结。

鉴于以上差异，在进行科技英语篇章的汉译时，应在尽可能保留原文篇章结构的基础上，根据汉语逻辑习惯对语序、句式等进行调整，以求实现最大对等，全面准确地传达原文的信息和逻辑。

科技文本的汉译通常以两种顺序进行谋篇构局：自然顺序和逻辑顺序。

（一）自然顺序

自然顺序包括时间顺序和空间顺序，即按照事件发展的时间先后、操作行为的程序步骤、地理空间的自然排布等客观因素的顺序进行翻译。

（1）时间顺序

时间顺序又细分为年代时间顺序和过程时间顺序。年代时间顺序就是以具体日期、年份、时间作为谋篇布局的标志；过程时间顺序多见于使用说明书、操作手册等文本类型，以操作的步骤顺序为叙述逻辑。例如：

1. Launched in 1977, the Voyager 2 has been on mission for more than 44 years. It first reached Jupiter in 1979, and then explored Saturn in 1981 following which it was programmed to extend its mission to the Uranian system in 1986.

旅行者 2 号于1977 年发射升空，现已在轨超 44 年。1979 年，旅行者 2 号首抵木星，又在1981 年抵达土星开展勘探活动。随后其任务版图继续扩张，于1986 年抵达天王星。

2. The deicing processes include <u>first</u> using heated freezing point depressant (FPD) fluid, <u>followed by</u> using thicker aqueous solutions with a lower freezing point.

除冰程序为：<u>首先</u>使用热防冰液溶液除冰，<u>然后</u>使用更为黏稠、冰点更低的溶液进行防冰。

不难看出，例 1 的翻译以四个年份节点为框架，引出相应事件，完全按照事情发生的时间顺序展开译文；例 2 则依照承接词"first""followed by"框定的流程顺序描写除冰过程中的两大步骤。

(2) 空间顺序

空间顺序指以几何空间或地理空间为逻辑顺序，分为概括空间顺序和明晰空间顺序。概括空间顺序即对空间位置信息的描述较为抽象笼统，如 between、inside、outward 等词限定出的位置关系；明晰空间顺序所用的描述词十分清晰明确，如 at a 20° angle、50cm above 等。翻译时应按照空间铺开顺序翻译，层层递进。例如：

1. We will describe a simple triode as a vacuum tube with three electrodes <u>inside</u> an evacuated glass envelope. <u>Right in the center</u> will be one electrode, the cathode. <u>Surrounding</u> the cathode and well spaced from it is the anode. <u>Between</u> the cathode and the anode, and usually <u>quite close to</u> cathode, is a wire screen that completely en-closes the cathode.

我们将一个简单的三极管描述为在真空玻璃外壳<u>内</u>装有三个电极的真空管。其<u>正中间</u>是一个电极，也就是阴极。<u>环绕着</u>阴极的是阳极，但二者间存在一定距离。<u>在阴极和阳极之间</u>，通常是<u>较为接近</u>阴极处，有一个线屏，将阴极完全围住。

2. <u>At the rear</u> of the wings and stabilizers are small moving sections that are <u>attached to</u> the fixed sections by hinges. The hinged part of the <u>vertical</u> stabilizer is called the rudder; the hinged part of the <u>horizontal</u> stabilizer is called the elevator; the <u>outboard</u> hinged part of the wing is called the aileron.

在机翼和安定面的<u>后缘</u>有一些活动部件，它们通过铰链<u>与</u>机身<u>相连</u>。<u>垂直</u>安定面上的部件叫作方向舵，<u>水平</u>安定面上的部件叫作升降舵，<u>舷外</u>部件叫作副翼。

例 1 中，对于三极管构造的描述从位于最中间的组成部分——阴极入手，而后通过阴极和其他部件的位置关系，自然地引出了对其他部分的介绍。译文遵照原文由位置描述词限定出的逻辑框架逐条叙述；例 2 中，先通过与"机翼和安定面"的位置关系总体交代这三个活动部件的位置，而后再分别通过和某一已知元件的位置关系，明晰这些活动部件具体都在哪些部位产生作用。

(二) 逻辑顺序

逻辑顺序即以原文中的逻辑为依据组织译文，科技文本中常见的逻辑顺序包括原因与结果、主次排序、类推、比较和对比、例证、图解等。

(1) 原因与结果

原因与结果即基于因果逻辑关系组织信息,在科技文本中出现的频率最高,其标志词有 because、as、since、as a result 等。在处理这类关系时要注意,英语习惯结果在前原因在后,中文则通常先说原因后说结果,进行翻译转换时要适当调整语序,以符合译入语习惯。除此之外,还需注意因果逻辑隐含在语义中的情况,有些句段并无明显的因果逻辑标志词,但上下句互为因果,译者在翻译时首先要能够判断出这种隐形的因果关系,并通过适当增加逻辑连接词使这一关系更为明晰。例如:

1. In addition, the galaxy likely halted its stellar production <u>because</u> it was starved of fresh, outside material needed to make new stars.

另外,<u>因为</u>缺乏制造新恒星所需的新鲜外部物质,该星系中可能无法再产生恒星。

2. Astronomers largely assume it is a nearby whirlpool of gas <u>because</u> no stars can be resolved.

<u>因为</u>那时人们还分辨不出恒星,天文学家大多认为它是一个附近的气体旋涡。

3. The end of the Cold War has opened up the skies over Russia and China to aircraft flying to and from Asian destinations, <u>but</u> traffic is restricted <u>by the lack of ground-based navigation aids.</u>

冷战结束之后,俄罗斯和中国对往返亚洲各国的飞机开放了空域,但<u>由于地面导航设备不足</u>,空中交通量受到限制。

4. <u>The friction of the air producing much heat</u>, the skin of spaceships is covered with high temperature resisting porcelain.

<u>由于与空气摩擦会产生大量的热</u>,飞船表面覆有耐高温陶瓷材料。

5. Capable of holding over 300 passengers and flying safely, this type of aeroplane is used by many airlines.

由于能装载300多名乘客且飞行安全,这种类型的飞机被许多航空公司采用。

以上五例中,例1和例2原文中包含明显因果逻辑连词"because",但表原因的分句都位于句尾,在翻译时调换顺序,提前翻译,使译文符合汉语语序;例3中无明显因果标志词,但理解句意后可知"空中交通量受到限制"是"地面导航设备不足",因此翻译 by 引导的介词短语加入因果关系连词"由于";例4和例5中前半句都是伴随状语成分,为主句补充原因,因此在翻译时都要增加连词"由于",显化因果关系,使译文更加条理清晰,逻辑通顺。

(2) 主次排序

主次排序即按照内容的重要程度组织信息。一般来说,这一逻辑顺序都有明显的标志词,如 first、second、most、least 等。在翻译按主次顺序展开文段时,只需遵循原文结构,用合适的承接词将译文串联起来,体现相应层次即可。例如:

Research shows that three factors are primarily responsible for excessive smoke. The most frequent cause is an air mixture that is too rich; that is, the fuel-air ratio is greater than 10:1. The second most frequent cause is oil in the fuel in concentrations greater than 60 cc per gallon. A third, and somewhat less frequent, cause is the speed of the engine itself.

研究表明,导致发动机排烟过量的主要原因有三:一是空气混合物过于丰富,即燃料-空气比大于 10:1,这是最常见的原因;二是燃料中油的浓度超过每加仑 60 毫升;三是引擎自身转速不妥,这一原因不大常见。

以上译例中,"the most frequent cause""the second most frequent cause""a third, and somewhat less frequent cause"如完全字对字翻译为"最常见原因""第二常见原因""第三种不那么常见的原因",会给人冗余赘述之感,因而灵活处理为"一是""二是""三是",再在分句末补充"不太常见"这一信息。这样一来,译文结构清晰、整齐,对三种并列原因读者也一目了然。再如:

First, until about 20 years ago, commercial air transport was dominated by US manufacturers, who sold their aircraft in US $. This generated an "aviation currency" not only used for original aircraft trading, but also for second-hand transactions, maintenance, repair and overhaul works, payments of airport fee and so on. Second, throughout the world kerosene has always been traded in US $.

首先,直到大约 20 年前,商业航空运输一直都由美国制造商主导,他们以美元出售飞机。由此产生的"航空货币",不仅用于原始飞机的交易,还用于二手交易、维修和大修工程、机场费用支付等。其次,在世界范围内煤油也一直以美元交易。

以上译例中有非常明显的主次标志词"first"和"second"将整段分成两个意群。在翻译时要按照原文语序,保留标志词的句首位置,明晰译文的逻辑层次。在处理"first"和"second"时,译者没有直译为"第一"和"第二",而是选用了能够明晰主次顺序的"首先"和"其次",逻辑顺序更为突出,也更符合汉语习惯。

(3) 类推

类推即将某一新概念和已知概念联系起来,通过描述二者之间的共性,以更通俗易懂的方式向读者展现这一新概念的原理。这样既使说理更具有说服性,又有助于读者理解。类推顺序有时通过显性逻辑词,如"be analogous to"表现,有时则隐含在句意当中。译者在翻译时应有意识地通过适当的关联词凸显这一逻辑。例如:

1. Sound waves are created by the compression of the molecules of air. The resulting wave motion is analogous to that created in water when a rock is thrown in a pond.

声波由空气分子压缩产生。此类波浪运动就如把一块石头扔进池塘时水中激起的波动。

2. It is unnatural in a large field to have only one shaft of wheat, and in the infinite Universe only one living world.

大片农田里只长一垄麦子是不正常的,无垠的宇宙只有一个有生命的世界(地球)同样反常。

以上两句中,例1有明显的类推标志"is analogous to"可直接译为"就如";而例2的类推逻辑则隐含在"and"连接的两个并列结构中,在翻译时要有意识地增词,用"同样"一词将二者间的共性显现出来。

(4) 比较和对比

比较显示事物间的共性,常以 likewise、similarly、in the same way 等词为标志;对比显示事物间的差异,常以 unlike、in contrast、on the other hand 为标志。

值得注意的是,比较和类比是不同的,类比表现的是两种基本不同的事物之间的共性,而比较则显示的是同一类别两种事物的共性。在翻译含有比较和对比逻辑的句段时,通常将被比较方(已知概念)提前翻译。以对比为例,例如:

1. Unlike gas turbines or micro-turbines, steam turbines do not directly consume fuel.

与燃气轮机或微型涡轮机不同,蒸汽涡轮机不直接消耗燃料。

2. The lack of blue, metal-poor clusters suggests that NGC 1277 never kept growing by gobbling up surrounding galaxies—unlike our own Milky Way, which continues to cannibalize galaxies that come too close.

缺少蓝色、金属含量低的星系团表明,不像我们的银河系会继续吞噬那些离我们过近的星系,NGC 1277 从未通过吞噬周围的星系来持续成长。

例1用来表述对比的短句位于句首,因此在翻译时顺序翻译,不做改动;而例2中,虽然包含对比逻辑的部分位于原文句末,但为了更符合汉语的思维逻辑,将这部分置于中心句前翻译。

四、科技翻译中的逻辑错误分析

逻辑,简单来说就是研究思维方式的科学。逻辑翻译学提出者阎德胜教授认为,逻辑规律由概念、判断、推理组成。而翻译作为一门实用科学,也应符合逻辑规律,应"根据上下文的逻辑分析采用一定的翻译技巧,处理原文有关的语言现象,以使汉译文概念明确、判断恰当、推理正确"。在科技英语的翻译实践中,由于忽视语篇内在逻辑而犯下令人啼笑皆非的谬误者不在少数。本节将从逻辑概念、逻辑判断及逻辑推理三方面入手,分析翻译实践中典型的逻辑错误案例。

(一) 逻辑概念

概念是逻辑规律最底层结构,在翻译中通常用于描述词汇范畴。科技英语具有普通词汇科技化、合成词汇专业化等特点,对词汇选择的准确度要求颇高,因此概念性错误十分常见,例如:

1. What exactly keeps the Earth and planets remaining in certain orbits around

the sun without going astray?

究竟是什么促使<u>地球和行星</u>总是在一定的轨道上环绕太阳运行而不远离太阳呢?

这一句中存在属种关系不清的问题,"行星"属于属概念,其包括地球,而"地球"则是种概念,二者不在同一概念层面,不能并列处理,应修改为"究竟是什么促使地球和其他行星……"

2. The <u>metal</u> is then cast into the resulting hollow mold.

然后,把金属<u>浇入</u>成型的空模。

这一句的问题在于没有根据语义对概念进行限制,理解句意可知,这句话描述的是浇铸过程,金属只有呈液体状态时才能被"浇入"模具中,因此应修改为"把金属熔液浇入成型的空模"。

3. Velocity changes if either <u>the speed</u> or the direction changes.

如果<u>速率</u>或方向发生改变,那物体的运动速度也会随之改变。

"speed"在物理概念中为"速率",与"速度"有着本质上的差别。速率是标量,有大小没有方向;速度是矢量,有大小有方向。此句在汉译时明显没有将这一词汇放在对应的科学范畴考察,而是按照普遍意义上的词义进行翻译,导致译文存在错误,非常不准确,应改为"速率或方向发生改变……"。

通过分析以上错误译例可知,为避免概念层面的错误,译者首先要认识到,科技英语中出现的许多通用词汇都有其特殊的含义,甚至同一个通用词汇在不同的专业领域其含义大不相同,在翻译时一定要慎重,除了勤查字典,还要根据上下文语境,对概念进行适当限制或概括,以保证译文的准确性;同时还要有意识地判断各概念间的关系,以及是否在同一逻辑层面。

(二) 逻辑判断

逻辑判断指基于逻辑思维对不同概念间关系的判断。科技文本客观严谨,小到词汇、大到篇章,各个语言层面都存在隐形逻辑关系,需要译者基于判断选择最贴切的译法。此外,科技文本中长句多,从句层层嵌套,修饰语多,因此正确判断各部分,特别是中心词之间的逻辑关系是翻译的关键。如果缺少有意识的逻辑判断,很可能会漏掉各概念间的逻辑联系,难免发生信息传递不准确、译文生涩难懂,不符合译入语表达习惯等问题,让读者一头雾水。例如:

1. Nonlinear equations indicate that a small force can have, <u>unpredictably</u>, either a small or a large effect.

非线性方程表明,一股细小的力可以产生<u>不可预测</u>的或大或小的影响。

译文将否定副词"unpredictably"直译为形容词修饰语,读着不自然,不通顺。从深层结构来看,"unpredictably"一词隐含着转折的逻辑,可以处理为一单独短句,凸显和"either a small or a large effect"的逻辑关系。修改后的译文为"非线性方程表明,一股细小的力产生的效果可大可小,但<u>究竟是大还是小则难以预料</u>"。

2. <u>Small in size</u>, this apparatus can lift a weight as heavy as 500 kilograms.

这一<u>小体积</u>装置可以举起 500 千克的重物。

原文中句首的状语从句其实隐含着让步的关系,是这一装置体积和能力间的一种对比,而译文中仅将状语简单转化为定语,完全没有体现这一层逻辑,应改为"<u>尽管</u>这个装置体积小,<u>可</u>它能举起 500 千克重的东西。"

3. Magnetic measurements give no information about the extent of the thermal area at depth. Volcanic rocks from inside the field were found to be non-magnetic.

测磁工作没能得出这个地热区的范围。人们发现热田内部的火山岩是非磁性的。

表面来看原文中的两句之间没有什么联系,但剖析语义可以发现,正是因为火山岩没有磁性,才使得测磁工作没有得到理想结果,第二句实为前一句的原因。这一逻辑关系在前后两句的时态上也有所体现,基于这一逻辑判断,译文应修改为"以前的调查表明,这个地热区内部的火山岩为非磁性岩石,因此测磁工作没能测出这个地热区的范围。"

在审校阶段,译者需要运用逻辑判断,从篇章整体的角度,对译文进行分析、辨别和断定,这样可以有效找出译文中存在的谬误。根据判断时所用到的知识类型,逻辑判断可以分为常识维度的逻辑判断和专业维度的逻辑判断,例如:

4. If the flexible leads of a portable electric appliance fray and <u>break</u>, and the wires then happen to touch the metal parts of the appliance, these parts can give a severe electric shock to anyone who touches them.

假使携带式电器的挠性引线磨损而<u>断裂</u>,而这些线碰巧与电器的金属零件相接触,那么谁碰到这些零件,谁就会触电。

单从语法结构上看,这一译文似乎没什么问题。但对照原文,我们会发现译者认为"leads"与"wire"是等同的。其实,这里的"wire"是因"leads"磨损而裸露出来的金属线,因此应将前半句修改为"假使携带式电器的挠性引线<u>磨损而裸露其中的金属线</u>"。

5. The alloy has strong <u>aging strengthening effect</u> and good performance at intermediate temperature.

合金有很强的<u>老化强化效果</u>和优良的中温特性。

这句话包含名为"aging strengthening effect"的材料领域专有名词,其准确释义为"<u>时效强化效应</u>"。

通过分析以上两例误译,我们可以感受到,在翻译实践中既需要对常识性知识有所了解,又要对翻译材料中的专业知识有所涉猎,才能对原文做出正确的逻辑判断,避免错译。

(三) 逻辑推理

逻辑推理是指由已知判断推出新判断的过程。英语篇章对所要探讨的新概念总是习惯先提出,开门见山,因此在翻译时一定要注意分析到底哪一方才是已知判断,即一句话可能是按照正常逻辑推理顺序组织的,也可能反向叙述,先提出新判断,后交代得出这一新判断的已知判断依据。如果推理顺序错误,那么译文的意思可能与原文有出入,例如:

Such approaches to policy change can also provide incentives for the fishing sector to invest in data collection, with the prospect of increased harvest possibilities if resources are shown to have the capacity to sustainably support higher catch levels.

若渔业具备持续进行高捕捞量的能力,那么这种推动政策改变的方法也能鼓励渔业部门对数据收集的投资,同时也能增加捕获量。

仔细分析此句结构可知,if 引导的条件状语从句是前半句的前提,而 "with the prospect of increased harvest possibilities" 是对 if 从句的补充,所以句子的推理顺序是 if 从句→with 结构→句子主干。上述译文没有清晰地表现出这种推理逻辑,且弄错了 with 修饰的成分,可以修改为"如果渔业资源可以维持高捕捞强度,捕获量就会上升,此时这些启动并开展政策改变的措施就可以推动渔业产业对数据收集进行投资"。

在科技文本翻译中,遇到像上述句子这样的从句以及限定成分较多的句段时,一定要首先确定好各部分间的逻辑推理顺序,再适当调整各部分语序,以形成逻辑合理的译文,避免误译。

五、如何提高翻译中的逻辑思维能力

既然逻辑思维在科技翻译中如此重要,那么在日常的学习生活中,如何有意识地训练和提高自己的逻辑思维能力呢?

首先要广泛阅读。可以选择难度适当的英语科普文章自主学习,在阅读时对其中的长难句进行精读,分析词与词、句与句、意群与意群之间的逻辑关系,有意识地训练分条缕析、理顺逻辑顺序的能力。长此以往的积累,不仅能提高逻辑分析能力,还能对不同领域的专业知识有所了解,对日后的翻译实践大有裨益。

其次在进行翻译练习时,要严格对照参考译文,分析译文是按怎样的逻辑框架谋篇布局的,其中叙事说理的逻辑顺序和自己的译文有何不同。只有认识到了差距及其产生的原因,才能进一步提升。

逻辑思维能力的提高不可能一蹴而就,但坚持不懈、有计划、有意识地加强阅读积累和翻译实践,就一定能使逻辑思维能力得到提升。

课后练习

请将下列句子译为中文，注意其中的逻辑关系。

1. Were there any inhabitants on the moon, they would see our earth reflecting the light of the sun, like a huge mirror hung in the sky.

2. They studied the strange equations that we now know describe that unescapable pocket of space we call a black hole and asked what they really represented.

3. We can study the motion of the projectile by watching the motion of its centre of gravity, at which the mass of the projectile is considered to be concentretrated.

4. The same ozone that helpfully blocks ultraviolet light in the stratosphere can seriously damage your respiratory system when it is at ground level—where it can be inhaled.

5. Up close, the galaxy would appear to be ablaze with bright, young, blue stars, but it looks red because its light has been stretched to longer (redder) wavelengths by the expansion of the universe.

6. This rule says that a stealth aircraft should be designed so that every detection system arrayed against it has roughly the same range. There is no point in building an airplane that is invisible to radar at five miles if optical sensors can see it at 10 miles.

7. For this kind of QMs, the entanglement source can be flexibly selected, including deterministic entanglement sources, while remaining the capability of multiplexed operations, and therefore should be more efficient for quantum repeater applications.

8. But while Captain Rogers knew what had apparently been transmitted by the plane on coded military frequencies, he was not aware that the aircraft had been exchanging routine flight instructions with the civilian control tower, broadcast on open radio channels.

9. But then in 2019, NASA's Neutron star Interior Composition Explorer (NICER), an X-ray telescope installed on the International Space Station two years prior, measured the size of a 1. 4-solar-mass neutron star called J0030, which is 1000 light-years from Earth, to be about 26 kilometers across.

10. Now, using NICER data, two independent teams have performed the same analysis for another neutron star, J0740, located 3000 light-years from Earth. The

results are surprising. With 2. 1 solar masses，J0740 is the most massive known neutron star—about 50% more massive than J0030. Yet the two are essentially the same size—the two teams arrive at 24. 8 or 27. 4 kilometers across for the former，with uncertainties of several kilometers.

实 践 篇

综合实践1 航空航天科普作品

科普作品以宣传普及科技知识为目的,具有科学性、文学性、通俗性和趣味性四大特点。其语言通俗易懂,既有科技文本特有的科学性和信息量,又兼具文学性和通俗性。科普作品体裁多样,常见的有科技通讯、科普小品文及科普著作等。

翻译科普作品时,要忠实于原文,力求准确传达原文的科学知识。对于文中专业术语的翻译,一定要多查相关专业领域的资料和工具书,准确把握术语在文中的含义,切忌望文生义。汉语表达要有文采,充分体现科普作品的文学性,选词要通俗易懂,照顾一般读者的阅读趣味。

翻译练习 1.1

How Long Is a Day on Saturn?

1. The answer was hiding in the planet's rings.

2. For decades, it was a nagging mystery—how long does a day last on Saturn?

3. Earth pirouettes around its axis once every 24 hours or so, while Jupiter spins comparatively briskly, once in roughly 9.8 Earth-hours[1]. And then there is Venus, a perplexingly sluggish spinner that takes 243 Earth-days to complete a full rotation.[2]

4. With Saturn, it turns out the answer rippled in plain view, in the planet's lustrous rings.

5. After reading small, spiraling waves in those bands, sculpted by oscillations from Saturn's gravity, scientists reported this month in the Astrophysical Journal that one Saturnian day is a mere 10 hours, 33 minutes and 38 seconds long, measured in Earth time.[3]

6. "The rings are not only beautiful, they're very diagnostic of what's going on inside the planet,"[4] said Linda Spilker, project scientist for NASA's Cassini mission, which studied Saturn for more than a decade.

7. Saturn has been stubbornly secretive about its days. Its buttery clouds don't bear helpful markings that scientists might use to track the planet's rotation, and they can't easily use its nearly vertical magnetic axis[5]—as they have for Jupiter's more off-kilter alignment—to gather clues about the planet's interior.[6]

8. Scientists long relied on others, ultimately misleading clues to figure out how fast the ringedworld turns. Not until the Cassini spacecraft swooped, flipped

and twirled through the Saturn system did scientists realize that the answer was outside the planet itself, etched into its icy rings.

9. As Saturn spins, its internal vibrations inscribe telltale signatures in its rings; studying those markings is now termed "kronoseismology,"⑦ from kronos, the Greek name for Saturn, and seismo, for quakes and vibrations.

10. In the same way that a bell rings and creates pressure waves⑧ that jiggle our eardrums, a spinning Saturn produces gravitational oscillations that herd particles in the rings into filaments⑨. The filaments form visible spiral patterns within the rings, revealing motions deep inside the planet that can be linked to its rotation speed.

11. As Cassini traversed the Saturn system, it used light from background stars, shining through the rings, to capture details of the embedded spirals. Then, a team led by scientists at the University of California, Santa Cruz, worked backward from the spirals to determine how quickly Saturn rotates. The exact answer, in Earth-hours: 10:33:38.

12. "This is a great story, a great picture to see⑩—Saturn kind of almost ringing," said Dr. Spilker, who was not involved in the study. "This interaction with gravity produces these little ripples that Cassini can see."

13. She anticipated that the rate of rotation⑪ would be refined once scientists better understood how the churning layers inside the planet affect those oscillations. But it's clear that one Saturnian day doesn't leave much time for long, luxurious naps or lazily gazing at the alien moons spangling its skies.

【参考译文】

土星上的一天有多久？

1. 答案就隐藏在土星的光环之中。

2. 几十年以来，这个谜团一直困扰着我们——土星上的一天有多久？

3. 地球大约每24小时绕地轴自转一圈，而木星的自转速度相对较快，大约每9.8小时自转一圈（以地球时间为单位）。而金星需要243天（以地球时间为单位）才能完成一次自转，其速度之慢让人困惑。

4. 而土星的自转速度，从它闪亮的光环中就可以很容易找到答案。

5. 科学家发现，土星上的引力所引发的震动，会在土星环上形成微小的螺旋状波纹。他们在本月《天体物理学》期刊上发表报告称，用地球时间来计算，土星上的一天只有10小时33分38秒。

6. "土星环不仅外表美观，借助它们还可以搞清楚土星内部的情况，"美国国家航空航天局卡西尼计划的项目科学家琳达·斯皮尔克说。该计划已经对土星进行了十多年的研究。

7. 土星的自转周期一直很神秘。它表面覆盖着的黄油状云团没有任何显著的标志可以帮助科学家追踪其自转周期。利用土星近乎垂直的磁轴来收集其内部的线索，对科学家来说也并非易事——而木星则因为磁轴更倾斜一些，更便于科学家进行相关研究。

8. 长期以来，科学家们计算光环环绕的土星的转动速度靠的是另外一些最终让人误入歧途的线索。直到"卡西尼号"飞船俯冲、翻转、旋转着穿过土星系统，他们才意识到答案其实不在土星内部，而在它的冰环中。

9. 土星自转时，其内部震动在土星环上留下了暴露其状态的明显特征；现在，研究这些标记的学科被称为"环形地震学"，这个词源自土星的希腊名称 kronos 和 seismo，意思是地震和震动。

10. 钟声响起时产生的压力波会让我们的耳膜震动。同样，土星自转时产生的引力震动会将土星环中的粒子聚集成暗条。这些暗条在环内形成明显的螺旋状纹路，暴露了行星深处的运动，而这些运动可能与其自转速度有关。

11. "卡西尼号"穿越土星系统时，利用背后恒星发出的光线照射土星环来捕捉土星那些自带的螺旋纹路的细节。此后，加州大学圣克鲁斯分校的科学家带领一个团队，从土星的螺旋纹路进行逆向研究，以确定其自转速度。他们得到的确切答案是：10 小时 33 分 38 秒（以地球时间为单位）。

12. 没有参与这项研究的斯皮尔克博士称："这是一项了不起的发现之旅，一幅美妙的景象——土星似乎响起钟声。这种与重力的相互作用产生的波纹，被"卡西尼号"观测到了。"

13. 她预计，当科学家们对土星内部的旋涡状层如何影响这些震动有更深入的了解时，土星自转速度的数据将会更加精确。但在土星上，一天太短，没有多少时间来享受漫长奢侈的小睡，也没有多少时间可以懒洋洋地盯着天空中闪闪发光的外星卫星发呆，这一点是确凿无疑的了。

【注释】

① "Earth-hour"指以地球时间单位来计算。翻译时可采用译文后加括号进行解释和说明。

② "a perplexingly sluggish spinner"此处直译不符合汉语的表达习惯，可以使用分句法，将原句子拆分，把这个名词短语译成一个短句，这样汉语更加通顺。即为：金星……其速度之慢让人困惑。

③ 这个句子比较长，在处理英语长句时，要注意句子各部分之间的关系，必要时进行拆分重组。具体到这句，首先找出句子的主干："科学家发现……他们在……发表报告称……"，再厘清句中各概念的关系：引力，震动和螺旋状波纹之间的关系，表达出这些关系"土星上的引力所引发的震动，会在土星环上形成微小的螺旋状波纹。"

④ "diagnostic"原意为"诊断的、判断的",在这里不能译成"诊断土星内部的情况",这样表达不符合汉语的行文习惯,应译为"搞清楚土星内部的情况"。在此句中,真正搞清楚土星内部状况的是人,而使用土星环是其方法,所以要根据逻辑添加"借助",这样译文更通顺、流畅。

⑤ "magnetic axis":磁轴。

⑥ 为了语意连贯,破折号中的内容放在最后翻译,同时需要将句意补充完整。

⑦ "kronoseismology":环形地震学。

⑧ "pressure waves":压力波。

⑨ "filament":暗条。

⑤、⑦、⑧、⑨为专业术语,在翻译专业术语时需要注意:优先使用权威的术语查询资源(术语网站、双语词典、专业技术人员等);专业术语的多种译法难以取舍时,可以查询相关论文文献,利用网络搜索引擎查询使用量;术语翻译还要结合上下文,尽量贴近原文。

⑩ "This is a great story, a great picture to see."需要将句意补充完整,具体翻译。

⑪ "the rate of rotation"是自转速度。"更加精确的"应该是"数据"而不是"自转速度",此处应注意搭配,将其补充为"自转速度的数据"。

翻译练习 1.2

The Moon Is a Hazardous Place to Live

1. If we get back to the lunar surface, astronauts will have to contend with much more than perilous rocket flights and the vacuum of space.

2. Despite NASA's successful Apollo landings, humans have spent very little time on the moon.

3. In total, 12 Apollo astronauts lived on the lunar surface for roughly 10 days, and traveled outside their lander for only 80 hours. With such a short sojourn, they caught merely a glimpse of the risks associated with survival there.

4. The prospect of humans staying on the moon for longer stretches has grown, with the Trump administration pledging an American return there by 2024, and China planning its first crewed trip to the lunar surface the following decade.[①] That will require work by scientists to further assess the challenges. Here are the most serious risks.

5. Beyond the obvious hazards that arise from a rocket flight, zero-gravity nausea and a risky landing, the moon itself can be deadly.

6. When the Apollo astronauts walked on the moon, the dust clung to their

spacesuits, scratched their visors and made their eyes water and their throats sore. ② It damaged the seals on the boxes of rock and soil that they brought back to Earth. ③ It even smelled like gunpowder inside the lander.

7. Lunar dust, which is composed of shards of silica, is fine like a powder but it cuts like glass④.

8. And that makes it toxic—so much so that Harrison Schmitt complained of "lunar dust hay fever" after his journey on Apollo 17. And yet, in total, Mr. Schmitt spent just 22 hours on the surface. Future astronauts might traverse the lunar landscape for much longer, giving them ample time to breathe in the deadly dust.

9. A recent study even suggests that prolonged exposure could lead to more serious effects, like bronchitis or cancer.

10. Then there is the moon's wrinkled surface—a topography that might be more threatening than rugged terrain on Earth.

11. "A lot of people think the moon is like a desert, but it's actually more like an ocean," Dr. Richard Scheuring, a flight surgeon at NASA, said. "It has a very undulating surface, like sea swells."

12. Throw in the intense contrast between light and dark—unlike anything we see here on Earth—and those changes in the terrain can play tricks on your eyes⑤.

13. When Pete Conrad and Alan Bean landed on the moon during the Apollo 12 mission, one of their tasks was to enter the 650-foot-wide Surveyor Crater. But as they skirted its rim, searching for the best path down, they informed Houston that the crater was far too steep⑥.

14. Topographic maps, however, revealed an easy⑦, 21 degree slope. The sharp shadows had fooled the astronauts.

15. That means the act of simply walking around on the moon might be perilous.

16. The moon might also be rippling. A recent study suggests that our lunar neighbor is technically active, with moonquakes as large as 5. 5-magnitude earthquakes. That is bad news for future lunar bases, which might be vulnerable to the shaking.

Sleepy moon men

17. Sleep may not come easy on the moon. Our body clocks are wound by light exposure, as day sweeps to night once every 24 hours⑧. But on the moon, that same shift occurs once every 28 days.

18. An astronaut's circadian rhythm is going to be quite baffled. And the

stakes are high. On Earth, sleep deprivation makes us fuzzy, but on the moon it is plain dangerous.

19. "Sleep is the biggest thing to protect for a crew member," Albert Holland, a NASA psychologist, said[9]. Without it, he added, astronauts might find themselves unable to face life-threatening emergencies.

20. When NASA sent the Curiosity rover hurtling to Mars in 2011, it packed a stow away: an instrument designed to measure the radiation throughout the trek.

21. The results confirmed that any astronauts beyond Earth's protective atmosphere and magnetic field would receive a hefty dose of radiation—one that is more than 200 times higher than levels on Earth. This is perhaps the biggest challenge.

22. Even aboard the International Space Station, which is so close to Earth that it is somewhat shielded, radiation levels are high—so high that when astronauts close their eyes, they often see flashes of light as cosmic rays pass through their optic nerve.

23. "It's an instantaneous reminder of the fact that those high-energy particles are going through your whole body—not just through your optic nerve," said Chris Hadfield, a Canadian astronaut who commanded the station in 2013. "And it's a stark reminder that you're no longer under the protective womb of the planet that nurtured us."[10]

24. Those cosmic rays have the power to charge through both strands of DNA, snapping them in two and causing cancer. In addition, a dose of long-ermradiation affects your cognitive abilities—resulting in dementia, memory deficits, anxiety, depression and impaired decision-making.

Small rocks, big blasts

25. Radiation is not the only killer from deep space. On Jan. 20, when stargazers across the Western Hemisphere watched a lunar eclipse, some saw a flash of light strike the moon's reddened surface. A small, fast-moving meteoroid had slammed into the moon.

26. Although the moon is a library of impact events, the flash was a reminder that collisions still occur today. Because the moon lacks an atmosphere, the strikes are hard and fast, posing a serioushazard for future astronauts.

27. Current spacesuits can block a 9 millimeter round, Dr. Scheuring said, "but a micrometeorite could be going a lot faster and it could be a lot bigger."

The lunar underground

28. A trip will be possible only if scientists find ways to mitigate these challenges. And although a lunar habitat may never provide the protective bubble we take for granted on Earth, there may be a variety of options for safer lunar living.

29. To protect against lunar dust, Dorit Donoviel, the director of NASA's Translational Research Institute for Space Health, said that any future lunar habitat would need filters and a strong airflow so that the astronauts would not breathe in the sharp silica. And Dr. Scheuring added that such a habitat would need a strong airlock where astronauts could clean off the dust before entering the main living area.

30. The habitat would also need to be equipped with lights to help counter the effects of sleep deprivation. Similar lights have already been installed on the space station, where individual sleeping compartments can be darkened to simulate night.

31. Just what that lunar habitat will look like remains a tough question, in part because radiation can penetrate most materials now used in space construction.

32. If living on the moon's surface for a sustained period is necessary, some experts have proposed covering shelters in a thick layer of water, because hydrogen works as a shield.

33. Another solution seems only natural: building living quarters underground.① The lava tubes carved by ancient volcanic activity could potentially be turned into spacious living spaces.

34. All of these hazards may seem worth it to anyone willing to be a moon dweller.

35. "Anything in life involves some risk," Mr. Hadfield said. "Often the more interesting and the more unusual the things you do with your life, the greater the inherent risk."

【参考译文】

月球是危险的居住地

1. 如果重返月球表面,航天员们要面对的远不止危险的火箭飞行和太空真空环境。

2. 虽然美国国家航空航天局"阿波罗号"已成功登月,但人类在月球表面停留的时间非常短暂。

3. 共有12名"阿波罗号"航天员在月球表面逗留大约10天,在着陆器外仅度过了 80 小时。由于停留短促,他们对在月球表面生存的风险只能说是略窥一斑。

4. 特朗普政府宣称美国将在 2024 年重返月球,而中国也计划在未来十年实现首次载人登月。这样,人类在月球上有望能驻留更长时间。这就需要科学家们对登月面临的挑战做进一步的评估。以下是几种最严重的风险。

5. 除了来自火箭飞行、零重力恶心和危险着陆的一些可预见风险外,月球本身也可能危及生命。

6. "阿波罗号"航天员在月球行走时,月尘吸附在宇航服上,划伤面罩,引起流眼泪、喉咙疼痛等症状。月尘还破坏了运回地球的月球岩石和土壤标本容器的密封材料。进入着陆器内部的月尘甚至闻起来有一股火药味。

7. 月尘的成分是二氧化硅颗粒,如粉末一样细,又像玻璃一样锐利。

8. 因此,月尘具有相当的毒性。哈里森·施密特完成"阿波罗 17 号"登月后诉苦说自己患上了月尘花粉症,而施密特仅仅在月球表面停留了 22 小时。将来,航天员穿行于月球各种地貌间的时间或许远远超出 22 小时,因此更有可能吸入致命的月尘。

9. 近期,一项研究甚至发现,长期接触月尘可能引起诸如支气管炎或癌症等更严重的后果。

10. 第二种严重风险则源于月球的褶皱表面——这种地形可能比地球上崎岖的地形更危险。

11. "很多人以为月球像沙漠,但实际上它更像海洋,"美国国家航空航天局航空军医理查德·舒尔林博士说,"月球表面起伏很大,如同海浪。"

12. 再加上强烈的明暗对比——地球上没有任何地方与此相似——月球上的地形变化会让人产生视觉偏差。

13. 皮特·康拉德和艾伦·宾恩乘"阿波罗 12 号"飞船登月的一项任务是进入650 英尺(198.12 米)宽的勘测者陨石坑。二人绕着陨石坑边缘,不断寻找最佳进坑路径,最终却向美国国家航空航天局休斯敦中心报告说,陨石坑太过陡峭,无法进入。

14. 但地形图显示斜坡平缓,坡度仅 21 度。是那些险峻的暗影让航天员做出了错误判断。

15. 这意味着仅仅在月球上行走都可能相当危险。

16. 月球表面也许还在不断地震动。最近一项研究表明,我们的月球邻居地壳构造活跃,月震强度可达 5.5 级地震的强度。这对未来在月球建立基地尤为不利,因为月震可能会使基地受损。

昏昏欲睡的月球航天员

17. 在月球上入睡可能不容易。在地球上,由于每 24 小时完成一次昼夜交替,依照光照时间长短,人体生物钟会做出相应的调节。但在月球,一次昼夜交替的周期为 28 天。

18. 登月航天员的昼夜节律会相当紊乱,因此风险极高。在地球上,睡眠不足我们会迷迷糊糊,但在月球,睡眠不足简直危险至极。

19. 美国国家航空航天局心理学家阿尔伯特·霍兰德说,对航天员来说,睡眠首当其冲,亟须保护。他说,失眠状态下,航天员们可能无法处理生死攸关的不测事件。

20. 2011 年,美国国家航空航天局"好奇号"登陆火星时,还暗中装载了一台仪器专门用来测量整个旅程中的辐射情况。

21. 结果表明,一旦失去地球大气层保护和磁场,航天员就会受到高剂量辐射,其强度是地球上的 200 多倍。这或许是最大的挑战。

22. 即使在离地球很近且有一定防护的国际空间站,辐射强度也相当大——大到航天员闭上双眼时,常常会看到宇宙射线穿过他们视神经时发出的闪光。

23. "这些闪光即时提醒我们,高能粒子正在穿过你的整个身体——而不仅仅只是视神经。"2013 年空间站主要负责人加拿大航天员克里斯·哈德菲尔德说。"这分明在发出警示:你已经脱离了地球母亲的护佑了。"

24. 这些宇宙射线能够冲破 DNA 的双链,将其分成两段并诱发癌症。此外,长期受辐射还会影响认知能力,导致痴呆、记忆缺陷、焦虑、抑郁和决策能力受损。

小石头 大爆炸

25. 辐射并不是外太空的唯一杀手。1 月 20 日,西半球的观星者观测月食时,有人看到了一道闪光冲撞发红的月球表面,那是一颗快速移动的小流星体撞上了月球。

26. 尽管月球发生过很多次碰撞事件,但这道闪光提醒我们,直至今日碰撞仍在发生。由于月球缺乏大气层,这些撞击猛烈而迅速,对未来登月的航天员们构成了严重威胁。

27. 目前的航天服可以阻挡 9 毫米大小的陨石,舒尔林博士说,"但微陨星的速度可能快很多,体积也大很多。"

月球地下居住区

28. 只有科学家们发现缓解这些威胁的办法,登月之旅才有可能实现。尽管月球栖息地可能永远不会提供类似地球那样的我们习以为常的保护层,但在月球上可能有多种更安全的生活选择。

29. 与美国国家航空航天局合作的空间健康转化研究所所长多里特·多诺维尔说,为了防止月尘,未来所有的月球栖息地都需要借助过滤器和强劲气流来防止航天员吸入尖锐的二氧化硅晶体。舒尔林博士还说,这样的栖息地需要安装一个坚固的气闸,这样,航天员就可以在进入主要生活区前清理掉身上的月尘。

30. 栖息地还需配备灯光,以帮助解决睡眠不足的问题。目前空间站已经安装了类似的灯光设备,每个睡眠舱都可以将灯光调暗,模拟夜晚环境。

31. 月球栖息地会是什么样子仍是一个难以回答的问题,部分原因在于目前用于太空建设的大多数材料,辐射都可以穿透。

32. 如果必须在月球表面生活一段时间,有些专家建议用一层厚厚的水覆盖居住地,因为氢可以起到防护作用。

33. 另一个解决方案近乎天然：将生活区建在地下。远古火山活动形成的熔岩管可以改造成宽敞的生活空间。

34. 对于愿意在月球居住的人来说，所有这些危险似乎都不算什么。

35. "生活处处有风险，"哈德菲尔德说，"通常，生活中越有趣、越非同寻常的事，其内在风险就越大。"

【注释】

① 此句为长难句，应厘清句子结构，层层翻译。"Trump administration pledging an American return"和"China planning its first crewed trip"为同级伴随状语，应优先翻译，之后再翻译主句。

② 此句为表达难点，首先确定"月尘"为句子主语，"clung""scratch""made"为接续发生的动词，翻译时应体现这些动词的内在逻辑。

③ 此句为表达难点，首先明确句子主干"it damaged the seals"，译为"月尘破坏了密封材料"，之后再层层补充修饰词"运回地球的月球岩石和土壤标本容器"。注意"rock""soil"应补充翻译为"月球岩石""土壤标本"，使句子更加通顺，方便读者理解。

④ 此句为理解难点，先查阅"月尘"相关的背景资料，可知"月尘"具有尖锐的特性。因此"it cuts like glass"是形容其"锐利"而非"易碎"。

⑤ "play tricks on"原意为"捉弄""开玩笑"，此处应联系上下文，译为"让人产生视觉偏差"。

⑥ 此处需查阅背景资料，补充翻译"Houston"为"美国国家航空航天局休斯敦中心"。

⑦ "easy"原意为"容易的""舒适的"，若直译则过于生硬，结合上下文，此处可译为"斜坡平缓"。

⑧ "wind（wound）"原意为"给（钟表等）上发条"，联系上下文，此处应为做出调整的意思；"Our body clocks are wound by light exposure"可译为"依照光照时间长短，人体生物钟会做出相应的调节"。

⑨ 此句为理解难点，句中"sleep is the biggest thing"的叙述对象是"crew member"，因此应译成"对航天员来说，睡眠首当其冲，亟须保护"，翻译时应明确各部分的内在联系，避免错译。

⑩ 此句为表达难点，"you're no longer under the protective womb of the planet that nurtured us"若直译为"你不再在养育我们的星球的保护下"，过于生硬，因此译为"你已经脱离了地球母亲的护佑了"，更符合汉语表达习惯。

⑪ 注意"only natural"的译法。"only"修饰"natural"，起加强语气的作用。此处是自然而然的意思，可译为"近乎天然"或"再自然不过了"。

翻译练习 1.3

History of Flight

1. From prehistoric times, humans have watched the flight of birds, and longed to imitate them, but lacked the power to do so.

2. Logic dictated that if the small muscles of birds can lift them into the air and sustain them, then the larger muscles of humans should be able to duplicate the feat.

3. No one knew about the intricate mesh of muscles, sinew, heart, breathing system, and devices not unlike wing flaps, variable-camber and spoilers of the modern airplane that enabled a bird to fly[①]. Still, thousands of years and countless lives were lost in attempts to fly like birds.

4. The identity of the first "bird-men" who fitted themselves with wings and leapt off of cliffs in an effort to fly are lost in time[②], but each failure gave those who wished to fly questions that needed to be answered.

5. Where had the wing flappers gone wrong? Philosophers, scientists, and inventors offered solutions, but no one could add wings to the human body and soar like a bird.

6. During the 1500s, Leonardo da Vinci filled pages of his notebooks with sketches of proposed flying machines, but most of his ideas were flawed because he clung to the idea of birdlike wings[③].

7. By 1655, mathematician, physicist, and inventor Robert Hooke concluded that the human body does not possess the strength to power artificial wings. He believed human flight would require some form of artificial propulsion.

8. The quest for human flight led some practitioners in another direction

9. In 1783, the first manned hot air balloon, crafted by Joseph and Etienne Montgolfier, flew for 23 minutes. Ten days later, Professor Jacques Charles flew the first gas balloon.

10. A madness for balloon flight captivated the public's imagination and for a time flying enthusiasts turned their expertise to the promise[④] of lighter-than-air flight. But for all its majesty in the air, the balloon was little more than a billowing heap of cloth capable of no more than a one-way, downwind journey.

11. Balloons solved the problem of lift, but that was only one of the problems of human flight. The ability to control speed and direction eluded balloonists.

12. The solution to that problem lay in a child's toy familiar to the East for 2,

000 years, but not introduced to the West until the 13th century—the kite⑤. The kites used by the Chinese for aerial observation, to test winds for sailing, as a signaling device, and as a toy, held many of the answers to lifting a heavier-than-air device into the air.

13. One of the men who believed the study of kites unlocked the secrets of winged flight was Sir George Cayley. Born in England 10 years before the Mongolfier balloon flight, Cayley spent his 84 years⑥ seeking to develop a heavier-thanair vehicle supported by kite-shaped wings.

14. The "Father of Aerial Navigation," Cayley discovered the basic principles on which the modern science of aeronautics is founded; built what is recognized as the first successful flying model; and tested the first full-size man-carrying airplane.

15. For the half-century after Cayley's death, countless scientists, flying enthusiasts, and inventors worked toward building a powered flying machine.

16. Men⑦, such as William Samuel Henson, who designed a huge monoplane that was propelled by a steam engine housed inside the fuselage, and Otto Lilienthal, who proved⑧ human flight in aircraft heavier than air was practical, worked toward the dream of powered flight.

17. A dream turned into reality by Wilbur and Orville Wright at Kitty Hawk, North Carolina, on December 17, 1903. The bicycle-building Wright brothers of Dayton, Ohio, had experimented for 4 years with kites, their own homemade wind tunnel, and different engines to power their biplane.

18. One of their great achievements in flight was proving the value of the scientific, rather than a build-it-and-see approach. Their biplane, The Flyer, combined inspired design and engineering with superior craftsmanship. By the afternoon of December 17th, the Wright brothers had flown a total of 98 seconds on four flights.

19. The age of flight had arrived.

【参考译文】

人类飞行史

1. 从史前时代起,人类就一直在观察鸟类飞行,并渴望效仿鸟类,可惜力不从心。

2. 按理说,如果鸟类的小肌肉群能使其升空并保持飞翔姿态,那么人类身上更大的肌肉群应该也能完成同样的壮举。

3. 一直以来，人类无从得知鸟类的肌肉、筋腱、心脏、呼吸系统，以及与现代飞机的襟翼、变弯度和扰流板相差无几的各种机体组织是如何精准协作，助鸟儿飞上天空的。但在效法鸟类飞翔的数千年中，无数人付出生命的代价。

4. 第一批"鸟人"曾为了飞起来，给自己安上翅膀，勇敢地跃下山崖。尽管他们的名字已经湮没在历史长河中，但每一次失败都为有飞翔梦想的人们提出了亟需解答的课题。

5. 这些前辈哪里出错了呢？虽然哲学家、科学家和发明家们都提出了个各种解决方案，但那时没人能给人体增添羽翼，像鸟一样直冲蓝天。

6. 16世纪，莱奥纳多·达·芬奇在笔记本上画满了想象中的飞行器草图，但因为固守扑翼构想，他的大多数构想都存在着缺陷。

7. 到了1655年，数学家、物理学家和发明家罗伯特·胡克得出结论：人体缺乏力量来挥动人造翅膀。他认为想要飞行，人类必须借助某种形式的人工推进力。

8. 对人类飞行的执着促使一些实干家就此转换了思路。

9. 1783年，约瑟夫和艾蒂安·蒙哥尔费埃精心设计的第一个载人热气球飞行了23分钟。10天后，雅克·查尔斯教授放飞了历史上第一个氢气球。

10. 这股对热气球飞行的狂热大大激发了人们的想象力。一时间，飞行爱好者们将聪明才智一股脑转向悬浮飞行，认为其前景光明。但除却其蔚为壮观的悬浮景象，热气球不过是一簇翻滚的布团，只能单程顺风飞行。

11. 热气球虽然解决了升力的问题，但人类飞行的艰难何止于此。热气球飞行家们无力控制飞行速度和方向。

12. 人类在风筝这种儿童玩具中找到了问题的答案。在东方，风筝已有两千年历史，但直到13世纪才被引入西方。中国人用风筝进行空中观测，为船只航行测定风向，用风筝传递信号，把风筝当成玩具，而风筝身上却蕴含着将重于空气的装置升空的诸多秘诀。

13. 一些人相信研究风筝能解开有翼飞行之谜，乔治·凯利爵士就是其中一员。凯利出生于英国，十岁时，蒙哥尔费的热气球升空飞行。在他84年的生命历程中，一直致力于研发比空气重、采用风筝状机翼的飞行器。

14. 作为"航空之父"，凯利的发现为现代航空科学奠定了基石，他还建造了公认的第一架成功的飞机模型，并对第一架全尺寸载人飞机进行了测试。

15. 在凯利去世后的半个世纪里，无数的科学家、飞行爱好者和发明家众志成城，一心想建造出动力飞行器。

16. 在实现动力飞行梦想的过程中，许多人都做出了贡献：威廉·塞缪尔·汉森设计出了一种大型单翼飞机，由机身内安置的一台蒸汽机驱动；奥托·李林塔尔印证了人类驾驶重于空气的飞行器飞行切实可行。

17. 1903年12月17日，威尔伯和奥维尔·莱特在美国北卡罗来纳州的基蒂霍克将千年梦想变为现实。莱特兄弟出生于俄亥俄州代顿市，本行是自行车制造。他们用了四年时间利用风筝、自制风洞进行实验,用不同的发动机为自己的双翼飞机提供动力。

18. 莱特兄弟在飞行方面的伟大成就之一是验证了科学方法的实用价值,而"造造看"的粗放尝试就相形见绌了。他们建造的双翼飞机"飞行者"1 号,融合了灵感十足的工程设计与高超绝伦的建造工艺。截至 1903 年 12 月 17 日下午,莱特兄弟进行了四次飞行,飞行时间总计 98 秒。

19. 人类飞行的时代终于来到了。

【注释】

① 此句为复杂句。muscles, sinew, heart, breathing system 与"devices"为并列成分,"not unlike wing flaps, variable-camber and spoilers of the modern airplane"用来修饰"devices"。

② who 引导的从句用来修饰先行词"bird-men",如果严格按照英语语序翻译将会显得修饰部分过于冗长,所以在翻译的时候应当打破原有的句式结构——先翻译 of 后面的定语从句部分,再翻译剩下的部分——以求译文的通顺流畅。同时,对"leapt off"的翻译增译为"勇敢地跃下山崖",更加贴合原文语境。

③ 达·芬奇最初绘制的为扑翼机草图。扑翼机是指机翼能像鸟和昆虫翅膀那样上下扑动的重于空气的航空器,又称振翼机。故此处将"birdlike wings"译为"扑翼构想"。

④ "promise"作名词时有"获得成功的迹象"之意,结合前文,此处将其单独提出来译为"认为其前景光明"。

⑤ 这句话是一个长句,翻译的时候不应囿于原有的句式结构。参考译文先翻译了句子的主干部分,同时将破折号后的"the kite"提到开头;而后再翻译"familiar …13th century"这部分的修饰成分。

⑥ 乔治·凯利(George Cayley)出生于 1773 年,于 1857 年去世,享年 84 岁;同时,文章第九段提到,蒙哥尔费是在 1783 年放飞了其热气球(即乔治·凯利 10 岁时)。为求流畅,译文灵活译出了这部分背景知识。

⑦ 这句话是一个长句,主句应为"Men worked toward the dream of powered flight.",中间插入的插入语对主语进行了举例说明。在翻译的时候应当注意调整语序结构,先翻译主句,再举例说明,从而更符合汉语习惯。

⑧ 奥托·李林塔尔于 1891 年制作了第一架固定翼滑翔机。他把自己悬挂在机翼上,从 15 米高的山冈上跃起,用身体的移动来控制飞行。

综合实践 2   航空航天科技说明书

科技说明书类文本语言简练、表达明确客观、不重文辞修饰,非人称句、公式化语句多,大量使用专业词汇、半专业词汇以及缩略语和符号。

　　翻译练习 2.1 是一篇科技产品说明书,属于呼唤性文本,即用客观、专业和简明的语言向读者传达产品信息。翻译时要遵循实用性原则,注意用语符合专业规范。翻译练习 2.2 涉及航空领域内容,专业术语多、用语简明正式,省略句多。翻译过程中如有疑问,需查找相关领域的资料或向专业技术人员咨询,务必做到透彻理解原文,以保证译文的可信度和标准度。文中专业术语的翻译,切忌生搬词典释义。汉语表达要注意用语简练、专业、明确和规范。

翻译练习 2.1

DJI Mavic 3 User Manual

　　1. Flight Modes: DJI has three flight modes, plus a fourth flight mode that the aircraft switches to in certain scenarios. Flight modes can be switched via the Flight Mode switch on the remote controller.

　　2. Sport Mode: In Sport Mode, the aircraft uses GNSS for positioning and the aircraft responses are optimized for agility and speed making it more responsive to control stick movements. Note that obstacle sensing is disabled and the maximum flight speed is 21 m/s(19 m/s when flying in the EU).

　　3. Cine Mode: Cine mode is based on Normal mode and the flight speed is limited, making the aircraft more stable during shooting.

　　4. The aircraft automatically changes to Attitude (ATTI) mode when the Vision Systems are unavailable or disabled and when the GNSS signal is weak or the compass experiences interference. In ATTI mode, the aircraft may be more easily affected by its surroundings. Environmental factors such as wind can result in horizontal shifting, which may present hazards, especially when flying in confined spaces.

　　5. The Forward, Backward, Lateral, and Upward Vision Systems[①] are disabled in Sport mode, which means the aircraft cannot sense obstacles on its route automatically.

　　6. The maximum speed and braking distance of the aircraft significantly increase in Sport mode. A minimum braking distance of 30 m is required in windless conditions.

　　7. A minimum braking distance of 10 m is required in windless conditions while the aircraft is ascending and descending.

　　8. The responsiveness of the aircraft significantly increases in Sport mode, which means a small control stick movement on the remote controller translates into the aircraft moving a large distance.[②] Make sure to maintain adequate maneuvering

space during flight.

9. DJI Mavic 3③ features both an Infrared Sensing System and Forward，Backward，Upward，Lateral and Downward Vision System，allowing for hovering and flying indoors as well as outdoors and for automatic Return to Home while avoiding obstacles in all directions. The aircraft has a maximum flight speed of 47 mph (75.6 kph) and a maximum flight time of 46 minutes. ④

10. The DJI RC Pro③ remote controller has a built-in 5.5 in high brightness 1000 cd/m2 screen with a resolution of 1920 ×1080 pixels. Users can connect to the internet via Wi-Fi while the Android operating system includes Bluetooth and GNSS. The DJI RC Pro③ has a maximum operating time of 3 hours. The RC-N1 ③remote controller displays the video transmission from the aircraft to DJI Fly③ on a mobile device. The aircraft and camera are easy to control using the onboard buttons and the remote controller has a runtime of 6 hours.

【参考译文】

大疆无人机用户手册

1. 飞行模式：大疆无人机支持三种飞行模式，外加飞行器在特定场景下切换到的第四种飞行模式。飞行模式可通过遥控器上的挡位切换开关进行切换。

2. 运动模式：使用全球导航卫星系统(GNSS)定位，飞行器响应的敏捷性和速度已优化，可以更快地对控制杆活动做出反应。注意：运动模式下，避障功能处于关闭状态，最大飞行速度为 21 米/秒(欧盟地区为 19 米/秒)。

3. 平稳模式：平稳模式在普通模式的基础上限制了最大飞行速度，使飞行器在拍摄过程中更加稳定。

4. 在 GNSS 卫星信号差或指南针受干扰并且视觉定位系统无法使用或出现故障时，飞行器将自动进入姿态(ATTI)模式。姿态模式下，飞行器更易受周围环境的影响。像风这样的环境因素会导致飞行器在水平方向产生漂移，特别是在狭窄空间飞行时，会危害飞行安全。

5. 运动模式时，无人机水平全向功能无法使用，即无人机无法自动识别航线上的障碍物。

6. 使用运动模式飞行时，飞行器的最快飞行速度和最长制动距离将大幅提升。在无风环境下，应预留至少 30 米的制动距离以保障飞行安全。

7. 在无风环境下，飞行器爬升或下降时，应预留至少 10 米的制动距离以保障飞行安全。

8. 使用运动模式飞行时，飞行器灵敏度大幅提升，具体表现为遥控器上小幅度的操作会导致飞行器产生大幅度的飞行动作。飞行时，应预留足够的飞行空间以保障飞行安全。

9. DJI Mavic 3 配备红外传感系统和水平全向、上视、下视视觉系统，能在室内外悬停和飞行，具备自动返航以及全向避障功能。飞行器最大飞行速度为 47 英里/小时(75.6 千米/小时)，最长续航飞行时间为 46 分钟。

10. DJI RC Pro 遥控器内置 5.5 英寸 1000 cd/m² 高亮显示屏，分辨率为 1920×1080 像素。用户可以通过 Wi-Fi 联网，安卓(Android)操作系统配备有蓝牙和全球导航卫星系统(GNSS)。DJI RC Pro 最长工作时间为 3 小时。RC-N1 遥控器可在移动设备上显示从飞行器到 DJI Fly 的视频传输。通过机载按钮可轻松控制飞行器和相机，遥控器的工作时间为 6 小时。

【注释】

① 无人机的前向、后向、侧向视觉系统覆盖水平方向全部视域，因此翻译时为保简洁，将这三部分并列结构概括总结为"水平全向"。

② 本句将"which means"灵活翻译成"具体表现为"，解释性更强。原文中"a small…movement"和"moving a large distance"的语法功能不同，在翻译时进行调整，选用统一的表述方式，"小幅度的操作"和"大幅度的飞行动作"对应整齐，更能凸显出一小一大之间的鲜明对比。

③ 对于 DJI RC Pro、DJI Fly 和 RC-N1 这类的产品型号，如果汉语中尚无固定译法，可直接引用原文，以免引起误解。

④ "a maximum flight time"译为"最长续航飞行时间"。

翻译练习 2.2

How Pilots Use Air Navigation to Fly

1. Air navigation is accomplished by various methods. The method or system that a pilot uses for navigating through today's airspace system will depend on the type of flight that will occur (VFR or IFR), which navigation systems are installed on the aircraft, and which navigation systems are available in a certain area.

2. Dead Reckoning and Pilotage

3. At the most simple level, navigation is accomplished through ideas known as dead reckoning and pilotage. Pilotage is a term that refers to the sole use of visual ground references.① The pilot identifies landmarks, such as rivers, towns, airports, and buildings and navigates among them. The trouble with pilotage is that, often, references aren't easily seen and can't be easily identified in low visibility conditions or if the pilot gets off track even slightly.② Therefore, the idea of dead reckoning was introduced.

4. Dead reckoning involves the use of visual checkpoints along with time and

distance calculations. The pilot chooses checkpoints that are easily seen from the air and also identified on the map and then calculates the time it will take to fly from one point to the next based on distance, airspeed, and wind calculations. A flight computer aids pilots in computing the time and distance calculations and the pilot typically uses a flight planning log to keep track of the calculations during flight.

5. Radio Navigation

6. With aircraft equipped with radio navigation aids (NAVAIDS), pilots can navigate more accurately than with dead reckoning alone. Radio NAVAIDS come in handy in low visibility conditions and act as a suitable backup method for general aviation pilots that prefer dead reckoning.[③] They are also more precise. Instead of flying from checkpoint to checkpoint, pilots can fly a straight line to a "fix" or an airport. Specific radio NAVAIDS are also required for IFR operations.

7. GPS

8. The global positioning system has become the most valuable method of navigation in the modern aviation world. GPS has proven to be tremendously reliable and precise and is probably the most common NAVAID in use today.

9. The global positioning system uses 24 U. S. Department of Defense satellites to provide precise location data, such as aircraft position, track, and speed to pilots. The GPS system uses triangulationto determine the aircraft's exact position over the earth. To be accurate, a GPS system must have the ability to gather data from at least three satellites for 2-D positioning, and four satellites for 3-D positioning.

10. GPS has become a preferred method of navigating due to the accuracy and ease of use. Though there are errors associated with GPS, they are rare. GPS systems can be used anywhere in the world, even in mountainous terrain, and they aren't prone to the errors of radio NAVAIDS, such as line-of-sight and electrical interference.

【参考译文】

飞行员如何使用空中导航飞行

1. 空中导航通过多种途径实现。在当今空域系统中确定航向时,飞行员使用的导航方法或导航系统取决于其采用的飞行类型(目视飞行规则(VFR)或仪表飞行规则(IFR))、飞机上安装了哪些导航系统,以及哪些导航系统在特定区域可用。

2. 航位推测法和引航

3. 最初,导航是通过航位推测法和引航来实现的。引航是一个术语,指仅依靠肉眼可见的地面参照物进行导航的方法。飞行员确定河流、城镇、机场和建筑物等地标,航行于其间。引航的困难在于,通常情况下,飞行员不易看到参照物,特别在能见度低或飞行员偏离(即使是稍微偏离)航道时,更不容易确定参照物。因此,航位推测法得到采用。

4. 航位推测法包括使用目视参照点以及时间和距离计算。飞行员选定从空中很容易看到的参照点,并在地图上确认其方位,然后根据距离、空速和风速计算出从一个点飞到下一个点所需的时间。飞行计算机协助飞行员计算时间和距离,飞行员通常使用飞行计划日志来记录飞行过程中计算出的时间和距离。

5. 无线电导航

6. 借助飞机配备的无线电导航设备(NAVAIDS),飞行员可以比仅靠航位推测法更准确地按航向飞行。无线电导航设备在能见度低时可以派上用场。所以对那些更倾向于使用航位推测法的通用航空飞行员来说,这种设备可成为一种适用的备选,而且它们也更精确。飞行员不用从一个参照点飞到另一个参照点,而是可以直线飞到一个"飞行定位"或一个机场。仪表飞行规则操作还需要特定的无线电导航设备。

7. 全球定位系统(GPS)

8. 全球定位系统已成为现代航空业最有价值的导航方式,其高度的可靠性和精确性已经得到证实。这种系统大概是当今最常用的导航系统。

9. 全球定位系统使用 24 颗美国国防部卫星,为飞行员提供精确的定位数据,如飞机位置、轨迹和速度。此系统使用三角测量法来确定飞机在地球上空的确切位置。准确地说,这个系统必须能够从至少三颗二维定位卫星和四颗三维定位卫星收集数据。

10. 全球定位系统因其精确便利已成为导航的首选方式。尽管与这种系统有关的误差也存在,但比较罕见。此系统可在全球各地使用,甚至山区,而且避免了无线电导航设备的缺陷,如视线干扰和电干扰。

【注释】

① 本句中的"pilotage""visual ground reference"均为航空专业词汇,查询专业词典后确定译为"引航""肉眼可见的地面参照物"。在翻译本句时,可将原文中宾语转换为方式状语,并增补"导航的方法"作为译文的宾语,使得对"引航"这一术语的定性更为清晰直观,提高译文可读性。

② 本句"references aren't easily seen…"为被动句,翻译时需按照汉语语言习惯转换为主动句,并添加相应动作主体"飞行员"作为主语。为使逻辑关系更为明晰,将 and 连接的"aren't easily seen"和"can't be easily identified"拆成两句,增补"特别……更……"凸显二者间的递进关系。

③ 本句较长,在翻译时断为两句,保证译文简洁达意及说理的明晰性,且句中实际隐含因果逻辑关系:阅读上文可知,能见度低时,航位推测法更为适用;而本句前半句说这一设备"在能见度低时可以派上用场",因此可以作为航位推测法的备选,翻译

时增补"所以",明确因果关系。

翻译练习 2.3

Introduction to Plane Maker[①]

1. Plane Maker is a program bundled with X-Plane[②] that lets users design their own aircraft. Using this software, nearly any aircraft imaginable can be built. Once all the physical specifications of the airplane have been entered (e. g. , weight, wing span, control deflections, engine power, airfoil sections, etc.), the X-Plane simulator will predict how that plane will fly in the real world; it will model the aircraft's performance just like it does for X-Plane's built-in aircraft.

2. Airplanes are saved in Plane Maker just as one would save a word processing document. These files are then opened and flown in the X-Plane simulator. Users can create a[③]. zip file of all the components of the airplane and distribute that ZIP on the Internet for others to fly. Planes created by others can also be downloaded and used in the simulator. The fan community X-Plane. org's "Download Manager" page is a good place both to upload and download these planes.

3. Performing a Test Flight

4. At some point—any point you choose, really—you will need to bring your aircraft into X-Plane and test it out. A typical test flight might include:

5. checking all control surface movements (by switching to an external view and giving full deflection in each direction for all controls),

6. noting the aircraft's ability to fly straight-and-level (how much trim it requires[④] , etc.).

7. confirming that the center of gravity feels like it is where it should be, and

8. checking the lift and drag vectors (by pressing the keyboard's / key by default) and confirming that they appear as expected.

9. Ideally, when performing a test flight, you will have enough experience with the real-world version of the aircraft to know how it should feel. Following a test flight, your goal is to be able to go back into Plane Maker and track down the source of any inaccuracies.

10. When modifying the flight model to match real-world performance, be careful that you do not modify your aircraft in a way that ignores the real-world meaning of the parameters you are changing. Even in the (unlikely) event that this is an acceptable fix at the moment, you risk degrading the quality of your aircraft in future versions of X-Plane.

11. For instance，suppose that，for some reason，your plane feels sluggish when turning. You might try increasing the area of the control surfaces，knowing that you had already matched the size of the real-world control surfaces，and find that the problem goes away.

12. This is absolutely not way to go about changing the plane's flight characteristics in X-Plane. The control surfaces' area does not simply represent a variable that can be changed to affect how a plane turns. The surface area must match the way the real-life airplane is built. If you increase the area，straying from reality，to "fix" a problem⑤，what you really do is create a new problem later when X-Plane goes to simulate your model.

13. Simply put，if you putintentional errors into your plane's flight model to compensate for limitations of the simulator，any future improvement in the simulation accuracy of X-Plane is almost guaranteed to make your plane fly worse in the future.

【参考译文】

Plane Maker 简介

1. Plane Maker 是一个随附在 X-Plane 里的、可以让用户设计自己飞机的程序。借助这个程序，用户可以建造几乎任何可以想象到的自己的飞机。只要输入所有的机身数据（如重量、翼展、控制偏差、发动机功率、机翼截面等），X-Plane 模拟器便能预测飞机在现实中的飞行情况；X-plane 模拟飞机性能，与模拟自己内置飞机如出一辙。

2. 用户可以像保存文字处理文档一样将飞机保存在 Plane Maker 中。随后只要在 X-Plane 模拟器中打开这些文件，飞机即可飞行。用户可以创建后缀为 .zip 的压缩文件，内含飞机的所有组件，并在互联网上共享该文件供他人飞行使用。用户也可以下载其他人创建的飞机，并在模拟器中使用。X-Plane.org 粉丝社区的"下载管理器"页面即可满足上传和下载需求。

3. 试飞

4. 无论选择什么时间点，用户都需要先在 X-Plane 中对飞机进行测试。一般的试飞包括以下步骤：

5. 检查操纵台的所有控制动作（切换到外部视图，并对所有操控杆在每一个方向都全幅度操作）。

6. 注意飞机平直飞行能力（飞机需要调整多少度才能保持平衡，等等）。

7. 确认飞机重心处在正确位置，以及

8. 检查升力和阻力矢量（默认通过按下键盘/键来完成此操作），并确认其符合预期。

9. 理想状态下,用户在试飞时能充分体验现实版的飞行器,从而了解飞行的真实感受。经过试飞,用户就能够重新回到 Plane Maker 软件中,对任何偏差进行溯源。

10. 在改进飞行模型以匹配其在现实中的性能时,注意不能以忽略所更改参数的真实含义为代价。即使目前这样做是可以接受的(虽然这种情况不太可能),但在未来版本的 X-Plane 中,存在飞机质量下降的风险。

11. 例如,假设飞机出于某种原因转弯迟缓(出现卡顿)。因为用户已经与真实操纵台的大小相匹配,这时可尝试增加操纵台的面积,这样问题就解决了。

12. 决不能在 X-Plane 中采用这种方法来改变飞机的飞行特性。操纵台面积并非简单代表一个变量,可以对其做出更改来改变飞机的转弯方式。操纵台面积必须与建造真飞机的方式相匹配。如果为了"解决"一个问题,不顾现实情况而增加其面积,那么后续 X-Plane 在模拟这个飞机模型时,这种做法实际上会造成新的问题。

13. 简单讲,如果把心知肚明的差错植入飞机模型来弥补模拟器的种种局限的话,那么将来对于 X-Plane 模拟精度做出的任何改进,几乎确定无疑地会让飞机的在模拟飞行中表现得更糟。

【注释】

① 本文为科技文本,在翻译的整体过程中要注意语言风格。一般情况下可以省略不必要的人称、连词等,以求语言简洁凝练,凸显较强的专业性。

② X-Plane 是一款模拟飞行软件。对于如 X-Plane 和 Plane Maker 这种暂无确切中文翻译的软件名称,可保留其英文形式,以避免误解。

③ 在科技文本的翻译中,诸如此处出现的"a"等冠词在保证译文通顺的前提下,可以省略不译,以求简洁明了。

④ 飞机配平(Aircraft trim)是利用装置对操作面(副翼、升降舵、方向舵)进行微调,来达到稳定飞机的姿态及航向的目的,这样可以降低飞行员调整或保持飞行姿态所需要的力量。依此背景知识,此处可翻译为"飞机需要调整多少度才能保持平衡"。

⑤ 这里三个短句以逗号连接,为求语意流畅连贯。此处在翻译时可以打破原有结构,调整句子顺序的同时进行整合,译为"如果为了'解决'一个问题,不顾现实情况而增加其面积"。

综合实践 3　航空航天科技报告

科技报告内容具体翔实、结构严谨,逻辑性强,加上句式结构复杂,专业词汇多,文体正式,没有专业背景的读者不易读懂。翻译这一类文本要求译者具有一定的专业基础知识,透彻理解原文。对于其中专业术语的翻译,要力求准确、简练、具体,切忌望文生义。译文要体现原文的科学性、逻辑性,用语要规范正式。

翻译练习 3.1

HIPPARCOS: STARS IN 3-D (489)[①]
NEW PUZZLE: KEY STARS ARE CLOSER THAN EXPECTED

1. Visible to the naked eye in the constellation of Taurus, the Pleiades have a special place in the science of the stars. Heavy stars burn up faster and grow old more rapidly than less massive stars. As all the stars of the Pleiades cluster formed at the same time, scientists can see how the mass of each star affects its luminosity, colour and ageing. Theories of stellar evolution, worked out using the Pleiades and other clusters of stars, under-pin all efforts to understand the Universe and to predict, for example, how long the Sun will survive.

2. But when Hipparcos measured the distance of the Pleiades, the theories[②] looked wrong. At 385 light-years, the star cluster is 10 percent closer than previously thought. The stars are 20 per cent less luminous - a huge discrepancy with the standard theories. Scientists reacted to the Hipparcos result with words like "amazement", "alarm" and even "nightmare"[③].

3. Hipparcos scientists can find no reason to think that the measurement is faulty, so the theorists are left scratching their heads about how to put matters right.

4. "Some colleagues are so unhappy about the unexpected distance of the Pleiades[④], they insist that Hipparcos is wrong," said Floor van Leeuwen of Cambridge University, UK. "After checking it in several ways, I'm sure Hipparcos gave the right answer. So don't use old theories to'correct' new data. Let's develop better theories."

5. Hipparcos was not conceived as an imaging telescope. It measured angles between points of light in the sky. But by combining repeated observations from different phases of the satellite's operating life, Swedish astronomers made the succession of images.

6. Two stars shift their positions as they orbit around each other. The precision is astonishing.

7. The stars are 34 light-years away in the Hydra constellation. They lie as close together as do the Sun and the planet Saturn. Their separation in the sky is only 1/5 000°. Yet the Hipparcos results show them plainly circling, and also moving together towards the east (left) in the course of two years.

8. Double stars are very useful in enabling astronomers to gauge the masses of stars. In this case the mass of the upper star⑤ is 0. 41 times the mass of the Sun, and that of its companion is 0. 42.

9. Hipparcos observed 24 000 double stars, of which 10 000 were not previously known to be double. The mission also investigated changes in luminosity of 8 000 new variable stars, in addition to 4 000 known variables - thus greatly improving the knowledge of how variable stars behave.

10. "Astronomers have always lived by their wits," said Lennart Lindegren of Lund Observatory, Sweden. "It was never easy to extract information from those small and distant lights in the sky. Now we're glad to show our colleagues some new tricks to make the Hipparcos results even more useful. "

【参考译文】

依巴谷卫星视角下的三维星空
谜题新现:主要恒星其实距地球更近

1. 在肉眼可见的金牛座中,昴星团在恒星科学中有着特殊的地位。比起小质量恒星,大质量恒星要燃烧得更快,衰老得更快。由于昴星团的所有恒星都是在同一时间形成的,科学家得以了解每颗恒星的质量对其亮度、颜色和衰老程度的影响方式。通过研究昴星团和其他星团而得出的种种恒星演化理论,全面助力人类认识宇宙,推动人类对太阳寿命等谜题做出预测。

2. 但是从依巴谷卫星测得的昴星团距离数据看,这些理论似乎存在偏差。依巴谷卫星测得的距离是 385 光年,比此前的数字要近 10%。测得的恒星亮度也比此前的数字低 20%,这一结果与标准理论推算出的数据相差甚远。科学家们用"惊愕""恐慌",甚至"噩梦"等字眼来形容他们对依巴谷卫星数据的感受。

3. 研究依巴谷卫星的科学家认为,测量数据绝无舛误,而那些理论派只能苦思冥想如何纠正偏差。

4."我的一些同事不相信这个出人意料的测量数据,坚持认为依巴谷卫星这次搞错了。"英国剑桥大学的弗洛·范·莱文说,"但在使用了几种方法对测量结果进行检验后,我确信数据准确无误。所以不要总是用旧理论来'纠正'新数据,而是应该提出更好的理论。"

5. 科学家并未把依巴谷卫星当成一台成像望远镜,而是用它来测量天空中光点之间的角度。但在卫星有效运行期间,瑞典天文学家通过整合不同阶段的多次重测数据,获得了连续的图像资料。

6. 这个星团中的两颗恒星在围绕彼此运行时会发生位移。依巴谷卫星对此的测量精度令人吃惊。

7. 这两颗恒星位于九头蛇星座，距离地球 34 光年。两恒星之间的距离很近，与太阳和土星之间的距离相当。它们在天空中的分离度只有 1/5 000°。然而，依巴谷卫星观测结果显示，它们明显地沿环行轨道运行，并且在两年期间结伴向东（左）移动。

8. 天文学家判定恒星质量时，双恒星作用巨大。而这次观测中，两恒星中靠上方的那颗恒星，其质量是太阳的 0.41 倍，而另一颗的质量则是太阳的 0.42 倍。

9. 依巴谷卫星一共观测到了 24 000 对双恒星，其中 10 000 对前所未闻。除了已知的 4 000 对变星外，依巴谷还观测到了 8 000 对新变星的亮度变化，极大提高了人类对变星运行特征的认知。

10. "天文学家一直依靠自己的领悟力生活，"瑞典隆德天文台的莱纳特·林德格伦说，"但从天空中依稀渺茫的光点中提取信息绝非易事。如今很高兴能向同事们展示一些新窍门，让依巴谷卫星的测量结果发挥更大作用。"

【注释】

① Hipparcos：依巴谷卫星（High Precision Parallax Collecting Satellite，缩写为 Hipparcos），全称为"依巴谷高精视差测量卫星"，是欧洲空间局发射的一颗高精视差测量卫星，以古希腊天文学家喜帕恰斯（又译作依巴谷）的名字命名，专门用于测量遥远星星的视差，从而计算距离。

② 测量恒星距离的白金标准是视差，此处 standard theories 并非回指上文的恒星演化论，而应译成"标准理论"。

③ 本句中三个形容词出现在"words like"后，用来形容科学家们对测量结果的震惊状态，这三个词均在引号内，意思层层递进。

④ 本句中出现的"the unexpected distance of the Pleiades"代指的是上文依巴谷卫星测量的昴星团距离，此处将其译为"这个出人意料的测量数据"，避免重复冗余。

⑤ 这里增译了"两恒星中"，限定了"the upper star"的范围，避免模糊不清。

翻译练习 3.2

NASA's Environmentally Responsible Aviation（ERA）Project

1. For over a century, NASA and its predecessor, the National Advisory Committee for Aeronautics（NACA）, have undertaken research and technology development that directly contributed to establishing air transportation as a corner-stone of our way of life. ① These efforts led to advances across the spectrum of study, from a basic understanding of the science of flight to practical engineering achievements that transformed concepts into reality. Although the shape and speed of commercial airliners have not changed significantly since the late 1950s, many

aspects of performance—such as range and fuel efficiency, and environmental impacts including noise and emissions—have improved tremendously. By the beginning of the 21st century, however, it became clear that more work was necessary. Heightened sensitivity to, and understanding of, the impact of aviation on the environment and the reduced availability of low-cost energy placed the spotlight directly on aircraft efficiency and reducing environmental impacts.② This realization spawned one of NASA's most ambitious aviation projects to date.③

2. NASA researchers believe that in the second quarter of this century④, commercial airline companies could save as much as \$250 billion thanks to so-called "green" aviation technology pioneered by the Agency and industry partners under NASA's Environmentally Responsible Aviation (ERA) project. This 6-year effort, which concluded in 2015, focused on development of technologies that will help aircraft manufacturers to reduce fuel consumption, exhaust emissions, and aircraft noise by increasing engine efficiency and improving overall aircraft design.

3. The Agency initiated the ERA project to explore, mature, and document the feasibility, benefits, and technical risks of vehicle concepts and enabling airframe and propulsion technologies originally identified in the Agency's Fundamental Aeronautics Program (FAP) and, in particular, the Subsonic Fixed Wing (SFW) project in order to mitigate the impact of aviation on the environment.⑤ NASA ultimately contributed more than \$400 million to the project, which also received approximately \$250 million from industry partners including Boeing and Pratt & Whitney.

4. The ERA project began in 2009 as part of the NASA Aeronautics Research Mission Directorate's Integrated Systems Research Program. Current-generation aircraft already benefit from NASA investments in aeronautical research of past decades. The development of digital fly-by-wire flight controls, supercritical airfoils, and winglets during the early 1970s and 1980s improved flying safety, controllability, and fuel efficiency, but most important, they became standard features of many modern aircraft. Once fully matured, technologies developed through the ERA project promise to become standard features of future generations of commercial air transports.

【参考译文】

美国国家航空航天局环境责任航空项目

1. 一个多世纪以来,美国国家航空航天局(NASA)及其前身美国国家航空咨询委员会(NACA)一直在从事研究和技术开发,其成果直接促成了航空运输成为人们生活方式的基石。从对飞行科学的基本理解,到将理论变成现实的应用工程成就,这些研发活动推动了整个研究领域的进步。自20世纪50年代后期以来,虽然商用客机在外观和速度上并没有重大变化,但许多方面的性能,如航程和燃油效率,以及包括噪声和排放在内的环境影响因素都得到了极大的改善。然而,到了21世纪初,人们清楚地认识到,还有许多工作需要做。随着公众对航空影响环境的敏感度和认识的提高,以及低成本能源供应的减少,飞机效率和减少对环境的影响成为关注的焦点。这种认识上的变化催生了美国国家航空航天局迄今为止最雄心勃勃的航空项目之一。

2. 美国国家航空航天局的研究人员认为,在2026－2050年的25年内,借助NASA及其行业伙伴共同创设的NASA环境责任航空(ERA)项目中的所谓"绿色"航空技术,商业航空领域可以节省多达2 500亿美元的费用。这项为期6年的项目于2015年结束,其重点在于开发技术来提高发动机效率,改进飞机总体设计,以此帮助飞机制造商降低油耗、减少废气排放和减弱飞机噪声。

3. 为了减轻航空对环境的影响,在NASA的基础航空计划(FAP)中,尤其是亚声速固定翼计划(SFW)中,已经确定了飞行器概念、特定功能机身及推进等技术。而启动这个环境责任项目就是为了探索、促进和记录以上技术的可行性、效益和技术风险。该项目最终得到NASA超4亿美元的资金支持,波音公司和普惠公司等行业合作伙伴也资助了约2.5亿美元。

4. 环境责任项目始于2009年,是NASA航空研究任务理事会的综合系统研究计划的一部分。当前这一代飞机已经从NASA过去几十年在航空研究方面的投资中获益。20世纪70年代和80年代早期,数字电子飞行控制、超临界机翼和小翼型技术的发展,提高了飞行的安全性、可控性和燃油效率,但最重要的是,它们成为许多现代机型的标配。通过环境责任项目开发的技术,一旦完全成熟,便有望成为未来几代商业航空运输的标配。

【注释】

① 本句较长,翻译时可将句子主干和从句从"that"处断开,根据句意增补"其成果"作为从句部分的主语,既保证了译文语法正确,又和前半句衔接流畅。句末的"our"一词在翻译时进行了概念外延,表达原文的实际意义,避免字对字死译。

② 本句主语过长,因此翻译时通过增译"随着"处理为伴随状语,将原文中的宾语作为译文的主语,突出主体。原文中"sensitivity"和"understanding"两词之前无主语,但通过上下文,不难看出指的是"公众"对此的敏感度和认识,在翻译时要补充出这一行为主体。

③ 从上句可知,人们对航空影响环境的认知逐步提高了,这是一个变化的过程,因此翻译本句时"realization"不能仅仅字对字译为"认识",而要体现出产生的变化这一动态过程,因此处理为"认识上的变化",更好地照应前句。

④ 翻译本句时要注意英汉两种语言在时间表达法上的差异。"the second quarter of this century"译成"本世纪的第二个 25 年"显然不符合汉语习惯,而译成具体的时间年份,给读者更直观清晰之感。

⑤ 本句中"vehicle concepts""enabling airframe""propulsion"等词均为专业术语,需要查专业词典确定准确译法。在处理这一长句时,为遵循汉语"先因后果""先概念后解释"的语言习惯,采取逆译法,首先通过目的状语铺陈原因,再表述主干信息(确定了哪些技术),最后另起一句补充项目开展目的这一次要信息。

翻译练习 3.3

Flight Control System

1. This chapter focuses on the flight control systems① a pilot uses to control the forces of flight and the aircraft's direction and attitude. It should be noted② that flight control systems and characteristics can vary greatly depending on the type of aircraft flown. The most basic flight control system designs are mechanical and date back to early aircraft. They operate with a collection of mechanical parts, such as rods, cables, pulleys, and sometimes chains to transmit the forces of the flight deck controls to the control surfaces.③ Mechanical flight control systems are still used today in small general and sport category aircraft where the aerodynamic forces are not excessive.

2. As aviation matured and aircraft designers learned more about aerodynamics, the industry produced larger and faster aircraft.④ Therefore, the aerodynamic forces acting upon the control surfaces increased exponentially. To make the control force required by pilots manageable, aircraft engineers designed more complex systems. At first, hydromechanical designs, consisting of a mechanical circuit and a hydraulic circuit, were used to reduce the complexity, weight, and limitations of mechanical flight controls systems.

3. As aircraft became more sophisticated, the control surfaces were actuated by electric motors, digital computers, or fiber optic cables. Called "fly-by-wire", this flight control system replaces the physical connection between pilot controls and the flight control surfaces with an electrical interface. In addition, in some large and fast aircraft, controls are boosted by hydraulically or electrically actuated systems. In both the fly-by-wire and boosted controls, the feel of the control reaction is fed back to the pilot by simulated means.

4. Current research at the National Aeronautics and Space Administration (NASA) Dryden Flight Research Center⑤ involves Intelligent Flight Control

Systems (IFCS). The goal of this project is to develop an adaptive neural network-based flight control system. Applied directly to flight control system feedback errors，IFCS provides adjustments to improve aircraft performance in normal flight，as well as with system failures. With IFCS，a pilot is able to maintain control and safely land an aircraft that has suffered a failure to a control surface or damage to the airframe.⑥ It also improves mission capability，increases the reliability and safety of flight，and eases the pilot workload.

5. Today's aircraft employ a variety of flight control systems. For example，some aircraft in the sport pilot category rely on weight-shift control to fly while balloons use a standard burn technique. Helicopters utilize a cyclic to tilt the rotor in the desired direction along with a collective to manipulate rotor pitch and anti-torque pedals to control yaw. ⑦

6. For additional information on flight control systems，refer to the appropriate handbook for information related to the flight control systems and characteristics of specific types of aircraft.

【参考译文】

飞行操纵系统

1. 本章的重点是飞行操纵系统。飞行员使用该系统来操纵飞行力以及飞机的方向和飞行姿态。应该注意的是,飞行操纵系统和特性会因为飞机类型不同而有很大的差异。最基本的飞行操纵系统设计是机械的,这可以追溯到早期的飞机。诸如操纵杆、钢缆、滑轮,有时还有链条这样的机械部件协同合作,将驾驶舱操纵装置的力传递到舵面。如今,机械飞行操纵系统仍用于不承载较大空气动力的小型通用和运动类飞机。

2. 随着航空业的日益成熟以及飞机设计师对空气动力学知识的深入了解,航空业生产出了体积更大,速度更快的飞机。因此,作用在舵面的空气动力呈指数增加。为了使飞行员更容易控制舵面,飞机工程师设计出了更复杂的系统。最初,由机械回路和液压回路组成的流体力学设计用于降低机械飞行操控系统的复杂性、重量及各种限制。

3. 随着飞机变得越来越复杂,舵面可以由电机、数字计算机或光纤电缆驱动。这种飞行操纵系统被称为"电传操纵"。它用电子接口取代了飞行员操纵装置和飞行舵面之间的物理连接。此外,在一些大型高速的飞机上,使用液压或电驱系统来增强操纵。不管是电传操纵还是增强操纵,操纵反应的感觉以模拟的方式反馈给飞行员。

4. 当前,美国国家航空航天局德莱顿飞行研究中心已经开始进行智能飞行操纵系统的研究。该研究的目标是开发一种基于自适应神经网络的飞行操纵系统。智能飞行操纵系统可直接修正飞行操纵系统的反馈误差以提高正常飞行时以及出现系统故障时的飞机性能。有了智能飞行操纵系统，即使在出现控制面故障或机身损坏的飞机上，飞行员也能够操控飞机并安全降落。该系统还能改善任务完成能力，提高飞行的可靠性及安全性，同时减少飞行员的工作量。

5. 今天的航空器使用各种各样的飞行操纵系统。例如，一些运动类航空器依靠重心转移来操纵飞行，而气球还是使用标准的燃烧技术。直升机利用周期变距操纵将旋翼倾斜到所需的方向，同时使用一个集体变距操纵来控制旋翼俯仰和反扭矩踏板来控制偏航。

6. 有关飞行控制系统的更多信息，请参阅相关手册，以了解有关飞行操纵系统和特定类型飞机特性的信息。

【注释】

1. 根据术语在线网站和航空专业词典，"flight control system"指"以航空器为被控对象的控制系统。分人工飞行控制（操纵）与自动飞行控制系统，主要是稳定和控制航空器的姿态和航迹运动"。本文主要内容是飞行员对飞机的控制，所以翻译成"飞行操纵系统"。

2. 本句"It should be noted that …"中"It"是形式主语，真正的主语是"that flight control systems and characteristics can vary greatly depending on the type of aircraft flown"。因为主语太长，为了保持句子的平衡，置于句末。翻译时"It should be noted …"部分可直接译成"应该注意的是"，放在句首，后面跟原句真正的主语。

3. "control surfaces"在专指飞机上的"操控面""控制面""控制表面"时，应译为"舵面"。这类词属于半专业词汇，要勤查字典，选择符合文章语境的专业译法。

4. 本句中的"larger and faster aircraft"在翻译时，可以增词，译为"体积更大、速度更快"，使用四字结构，这样的汉语表述更明确，且体现译入语语言的形式美和语音美，充分发挥了汉语的表达优势。

5. 本句中的"National Aeronautics and Space Administration（NASA）Dryden Flight Research Center"（美国国家航空航天局德莱顿飞行研究中心）为专有名词，已有约定俗成的规范译法，勤查字典即可。文中第一次出现机构名称时，需全译加英语缩写形式，文中再次出现时，可以只用英语缩写形式即可。值得注意的是，原名称中并没有"美国"这个信息，但为了有些不了解美国国家航空航天局的读者考虑，可以加上"美国"。

6. 本句中"With IFCS"作状语，介词 with 可以动词化，译为"有了 "；"that"引导的定语从句"that has suffered a failure to a control surface or damage to the airframe"修饰"an aircraft"，这部分隐含逻辑关系，汉译时可以增加"即使"和"也能"，使逻辑关系显化。

7. 本句中的"cyclic"（周期变距操纵）和"collective"（集体变距操纵）都是航空专业术语，需查专业字典和资料解决，不能当作普通词汇随意翻译。

翻译练习 3.4

Hubble Space Telescope Celebrates 30 Years of Discoveries and Awe-Inspiring Images[①]

1. The Hubble Space Telescope launched 30 years ago on Friday, forever changing the way we see the universe. The telescope's ethereal, dreamy and almost fantasy-like views of space vistas have inspired people for decades and led to some of the most important astronomical discoveries. [②]

2. The space observatory and its instruments, an international cooperative effort between NASA and the European Space Agency, captures unprecedented views of stars, galaxies and the distant universe in visible, ultraviolet and near-infrared light. These different wavelengths of light have allowed Hubble to peer into different regions of space that had never been observed before.

3. It orbits the Earth from a distance of 340 miles, well above the distorting effects of Earth's atmosphere for observing space both near and far. [③] "This has given us a new vantage point for viewing everything in the universe from the nearby solar system to distant galaxies and opened our eyes to the richness of the content of the universe and dynamic activity of the universe over time. "

4. Hubble's work led to the revelation that our galaxy was one of many, forever changing our perspective and place in the universe. Hubble continued his work and discovered that distant galaxies appeared to be moving rapidly, suggesting that we live in an expanding universe that started with a big bang.

5. Over 30 years, Hubble has enabled astronomers around the world to study black holes, mysterious dark energy, distant galaxies and galactic mergers. It has observed planets outside of our solar system and where they form around stars, star formation and death, and it's even spotted previously unknown moons around Pluto.

6. Hubble has characterized the atmospheres of exoplanets and spotted weather shifts on planets in our own solar system. And it's looked across 97% of the universe, effectively peering back in time. The telescope was expected to last for 15 years, and it's still going strong. But Hubble was also designed to be serviced and upgraded over time.

7. Between December 1993 and May 2009, astronauts launched on the space shuttle and rendezvoused with the telescope to make repairs and replace gyroscopes and instruments. The first one, in 1993, helped fix Hubble's infamous mirror flaw that was causing blurry images to be returned by the telescope. Astronauts

installed corrective optics and new instruments to fix it.

8. Each mission, which took years of planning and preparation, required the astronauts to leave the shuttle and conduct spacewalks to and inside a component of the telescope for repairs and installing instruments. All while the telescope moved at 17,000 miles per hour at an inclined 28.5 degrees to the equator around the Earth.

9. Outside of our solar system, Hubble has explored our Milky Way galaxy and neighboring galaxies. The dramatic, colorful images Hubble is known for are largely of active nebulae in our galaxy, bright clouds of gas and dust where stars are forming. ④ "When Hubble was launched (in 1990), no one knew about a single planet outside of our solar system," said Tom Brown, the Hubble Mission Head at the Space Telescope Science Institute in Maryland. ⑤

10. Astronomers found exoplanets in the 1990s using other telescopes, but Hubble was able to do groundbreaking science by following up on those observations and study exoplanet atmospheres. Hubble also enabled astronomers to realize that galaxies tend to merge with one another, capturing dramatic images of these mergers unfolding across the universe. That's how our own Milky Way galaxy grew to its current size, through merging with smaller galaxies.

11. Hubble's scientists believe that the telescope will keep operating through at least 2025, if not longer. This provides astronomers with an excellent opportunity because Hubble can overlap with new space-based telescopes coming online soon, like NASA's James Webb Space Telescope set to launch in 2021.

12. But when Hubble's mission does come to an end, its optical and ultraviolet capabilities won't be reproduced anytime soon. Depending on when Hubble concludes, this could leave a massive gap for scientists who depend on Hubble's observations to do their work. ⑥

13. For now, they have hope that the telescope will continue on for years, and maybe even decades, to come. "It's aging in a very graceful, well understood way and operating just as powerful as ever," Brown said.

【参考译文】

哈勃望远镜：三十载科学探索，大宇宙奇观尽览

1. 30 年前的一个周五，哈勃太空望远镜发射成功，从此人类观察宇宙的方式彻底改变。哈勃望远镜下那缥缈、如梦似幻的太空美景几十年来一直启发着人类思考、想象，并促成了一些重大的天文发现。

2. 哈勃望远镜这座太空天文台及其搭载的各种仪器是美国国家航空航天局与欧洲航天局合作的结晶,可通过观测可见光、紫外光和近红外光,捕捉恒星、星系以及遥远宇宙的前所未有的图像。通过观测这些不同波长的光,哈勃望远镜能够观测到人类之前从未观测到的宇宙中的不同区域。

3. 哈勃望远镜在距地面 340 英里(约 547 公里)的高度绕地球飞行,这样远远超出地球大气层的高度,规避了大气层对深空和近空观测造成的畸变效应。"不论是我们近旁的太阳系,还是遥远的其他星系,哈勃为我们提供了全新的有利观测位置,帮助我们看到丰富的宇宙万物及宇宙中随时而动的各种动态活动,让我们眼界大开"。

4. 哈勃望远镜所拍摄的照片使人们意识到我们所在的银河系只是众多星系之一。这一发现彻底改变了人类的视野,也改变了我们的星球在宇宙中的定位。哈勃望远镜不断乘胜追击,又发现遥远的星系似乎在快速移动,这表明我们生活在以大爆炸为起点、不断膨胀的宇宙中。

5. 30 年来,哈勃望远镜帮助世界各地的天文学家研究黑洞、神秘的暗能量、遥远的星系以及星系的合并。它还观测到了太阳系以外的行星以及这些行星围绕恒星附近形成的位置,也观测到恒星的形成与衰亡,甚至还发现了冥王星周围前所未知的卫星。

6. 哈勃望远镜描绘出了系外行星的大气层特征,并发现了太阳系各行星的天气变化。它的观测范围覆盖了 97% 的宇宙,成功观测到了宇宙过去的模样。哈勃望远镜预计还能再使用 15 年,至今仍运转良好。不过按最初设计,哈勃望远镜会不断得到维护升级。

7. 从 1993 年 12 月到 2009 年 5 月,航天员乘坐航天飞机多次与望远镜对接,对其进行维修并更换了陀螺仪和一些仪器。第一次维修是在 1993 年,这次任务修复了哈勃望远镜臭名昭著的镜面缺陷,从望远镜传回的图像模糊不清就是镜面缺陷造成的。航天员为此安装了校正光学器件以及其他新仪器。

8. 每次维修任务都需数年的计划与准备。执行任务时,航天员需要离开航天飞机,在太空中行走至望远镜,并置身望远镜部件内部来进行维修和设备更换。而在整个维修过程中,哈勃望远镜一直以 17 000 英里/小时(约 27 000 千米/小时)的速度,在与地球赤道面倾角为 28.5 度的轨道上绕地球旋转。

9. 除了太阳系,哈勃望远镜也对银河系和邻近星系进行了观测。那些哈勃望远镜拍摄的著名的梦幻般的多彩图像大多来自银河系中的活跃星云。星云是宇宙中的气体与尘埃结合成的明亮的云雾状天体,通常也是恒星形成的区域。美国马里兰州太空望远镜科学研究所负责人汤姆·布朗说:"1990 年哈勃望远镜发射之前,没人了解太阳系之外的任何行星。"

10. 20 世纪 90 年代,天文学家们用其他望远镜也发现了系外行星,但是哈勃望远镜能够在追踪观测的基础上,对系外行星的大气层进行研究,实现了科学上的突破。哈勃望远镜拍摄到宇宙中星系融合的壮丽场景,使天文学家们意识到星系之间存在着相互合并的倾向。银河系正是通过与较小星系的合并，才扩张到了现在的规模。

11. 负责哈勃任务的科学家认为，哈勃望远镜至少能运转到 2025 年，甚至更久。这为天文学家们提供了一个绝佳的机会，因为届时哈勃可以与即将投入使用的新型太空望远镜，如美国国家航空航天局将于 2021 年发射的詹姆斯·韦伯太空望远镜，搭配工作。

12. 但当哈勃确实要退役时，其光学和紫外线观测功能短期内无法得到替补。由于无法确定哈勃望远镜何时退役，那些需要哈勃观测结果来进行研究的科学家们可能会面临巨大的数据缺口。

13. 目前，这些科学家们希望哈勃能够继续使用数年甚至数十年。布朗说："哈勃望远镜正在非常优雅地老化，这再清楚不过了，但目前它的性能强劲，一如既往。"

【注释】

① 本文为科技报告。其标题语言精练，重点突出，翻译时在用语和修辞方面多费心思，既要语义上贴合原文，又要突出汉语文章标题凝练、上口的特征。依此翻译原则，本文题目可以处理成"哈勃望远镜：三十载科学探索，大宇宙奇观尽览"，符合译入语文章标题的特征，有助于吸引更多的读者。

② 本句中的名词短语"ethereal, dreamy and almost fantasy-like views of space vistas"（那缥缈、如梦似幻的太空美景），中心词"views"前面有三个形容词连用的前置定语，后面是 of 引导的介词短语作后置定语。翻译时不需要字字对应，可以按照汉语表达习惯，灵活处理，将中心意思译出即可。

③ 本句宜采用解释性译法，即不要局限于原文的语言形式，只要将原文想要表达的信息通过解释的方法用汉语说清楚、讲明白即可。例如句中"well above the distorting effects of Earth's atmosphere"可以解释性地译为"远远超出地球大气层的高度，规避了大气层对深空和近空观测造成的畸变效应"。

④ 本句虽然不是很长，但信息密集度高，翻译时需要断句，分成两部分，将句子中的各层含义一一表达出来。

⑤ 本句中有一个同位语，具体说明 Tom Brown 的身份。处理这种对前面的名词进行解释和补充说明的同位语时，可以考虑使用以下两种翻译方式：a.同位语不是很长时，可以直接译成名词短语，放在名词（人名）之前，作前置定语，如本句的处理方式"美国马里兰州太空望远镜科学研究所（STScl）负责人汤姆·布朗"；b.同位语较长时，可以和原句断开，单独译为一个判断句，并根据情况，放在原句之前或之后。以"Byrne said that usually the jet stream, a core of strong winds around six miles above the earth's surface, provides a combination of high and low pressure systems resulting in some rain over northern Europe."为例，"a core of strong winds around six miles above the earth's surface"是"jet stream"的同位语，比较长，翻译时如果处理成"jet stream"（高空急流）的前置定语，整个句子会显得层次不分明，意思不连贯，

翻译腔较浓,所以最好单独翻译成判断句,放在句首。此句可以翻译成"据拜恩说高空急流是地表以上 6 英里(9.6 千米)处的强风核心。通常情况下,高低压在此交锋,导致北欧出现降雨。"

⑥ 本句中的"Depending on when Hubble concludes"为非谓语动词形式,作伴随状语,可以解释性翻译成"由于无法确定哈勃望远镜何时退役"。

翻译练习 3.5

The Surface Search for MH370

1. The surface search for MH370 lasted from 8 March 2014 until 28 April 2014 and was initially a search and rescue operation. The intent was to locate the aircraft as quickly as possible in order to rescue any potential survivors. The areas searched were based on information from a range of sources and progressively refined analysis in relation to the aircraft's most likely flight path. ①

2. Early in the surface search the Malaysian Government convened the Joint Investigation Team (JIT) comprising experts from the People's Republic of China, France, Malaysia, United Kingdom, United States and Malaysian Government officials. Soon after, a satellite communications working group (SATCOM WG) was also formed and included experts from Inmarsat and Thales. These groups of experts worked together to provide advice to the Malaysian Government on the surface search areas.

3. By the end of the surface search an area of several million square kilometers had been searched by aircraft and surface vessels in the South China Sea, Andaman Sea, Bay of Bengal and Indian Ocean, however no items of debris from MH370 were recovered or positively identified.

4. The following section sets out the chronology of the surface search, the search assets used, the areas of focus at different times and the information or analysis used to define each area.

Malaysian led surface search②

5. Search operations commenced on the day that MH370 went missing, 8 March 2014, and were led by Malaysian and Indonesian authorities in areas around Malaysia until 23 March 2014.

East of the Malay Peninsula

6. Initially search and rescue (SAR) operations were coordinated by Kuala Lumpur Aeronautical Rescue Coordination Centre (KL ARCC) and were conducted to the east of the Malay Peninsula in the South China Sea between 8 and 15 March

2014. This area was based largely on the last contact with the aircraft and where the SSR transponders ceased to operate.

West of the Malay Peninsula

7. From 8 to 15 March 2014, the same time as the surface search in the South China Sea east of the Malay Peninsula, SAR operations coordinated by the Royal Malaysia Air Force, were conducted to the west of the Malay Peninsula.

8. These search operations were undertaken in the Strait of Malacca, Andaman Sea, Bay of Bengal and west of Sumatra from 8 to 15 March 2014. This area was searched based on the primary radar data which indicated an unidentified aircraft had flown up the Strait of Malacca, thought to be MH370. [3]

Northern and southern corridors announced

9. In the first week of the surface search, an analysis of Inmarsat satellite communication (SATCOM) data for MH370 indicated that the aircraft had flown for a further six hours after the final radar capture at the northern tip of Sumatra. The initial satellite data analysis indicated that the aircraft had flown along one of two corridors; one to the north in the direction of Kazakhstan or one to the south to the Indian Ocean. This new information led to the suspension of SAR operations to the east and west of the Malay Peninsula on 15 March 2014.

10. An aerial search coordinated by KL ARCC and Baden SAR National (BASARNAS), the Republic of Indonesia National Search and Rescue Agency, was conducted within the southern corridor indicated by the SATCOM data from 18 to 23 March 2014 .

Australian led surface search

11. On 17 March 2014, with the southern corridor extending into the Indian Ocean and the Australian search and rescue region, Australia assumed responsibility for coordinating the SAR operation for the aircraft, at Malaysia's request. [4] The Australian Maritime Safety Authority (AMSA) was responsible for coordinating this activity using aircraft and surface vessels operating from Western Australia until 28 April 2014. [5] AMSA continued to take advice on the areas to search from the JIT and SATCOM WG based on the progressive analysis of the Inmarsat satellite communication logs and other aircraft performance analysis.

12. The surface search was focused on the identification and recovery of any debris from the aircraft floating on the sea surface. When AMSA took over the coordination of the surface search, nine days had passed since the aircraft went missing. It was therefore necessary to define the areas to be searched by aircraft and surface vessels based on the analysis indicating where the aircraft may have

ended the flight and the calculated drift of a range of possible types of floating debris in the days after 8 March 2014. ⑥

13. A drift modelling working group was set up by AMSA，comprising a number of organizations including：CSIRO，Asia-Pacific Applied Science Associates，the United States Coastguard，the Bureau of Meteorology and Global Environmental Modelling and Monitoring Systems to ensure that best practice drift modelling was put in place for the surface search. ⑦ The drift modelling was also informed by the deployment of self-locating datum marker buoys（SLDMB）from aircraft and vessels throughout the surface search. A SLDMB is a drifting surface buoy fitted with a GPS that is used to measure surface ocean currents. The marker buoy has an expected lifetime of over 20 days once deployed in the ocean. Similarly，real-time wind and wave data from the search area was used to continuously update the drift models.

【参考译文】

海面搜寻马航 370 客机

1. 马航 370 的海面搜寻行动于 2014 年 3 月 8 日至 4 月 28 日进行。这场搜寻行动的最初意图是搜救，即尽快找到飞机，营救可能的幸存者。搜寻的区域是根据从各种来源获得的信息以及依照该飞机最有可能的飞行路线随时细化的分析来确定的。

2. 海面搜寻初期，马来西亚政府组建了联合调查组，成员包括来自中国、法国、马来西亚、英国和美国的专家以及马来西亚政府官员。不久，他们又成立了卫星通信工作组，其中包括来自国际海事卫星组织和法国泰雷兹集团的专家。这些专家小组通力合作，就海面搜寻区域向马来西亚政府提供建议。

3. 海面搜寻结束时，飞机和水面船舶已经搜索了数百万平方千米的区域，其中包括中国南海、安达曼海、孟加拉湾和印度洋，但没有找到或确认任何马航 370 客机的残骸。

4. 以下是此次海面搜寻的时间顺序、使用的搜寻物资和不同时间的重点搜寻区域以及用来确定每一搜索区域的信息和分析资料。

马来西亚主导的海面搜寻

5. 2014 年 3 月 8 日马航 370 客机失踪当天搜寻行动即开始，由马来西亚和印度尼西亚政府主导。此阶段的搜寻主要集中在马来西亚周边地区，于 2014 年 3 月 23 日结束。

马来半岛东部的情况

6. 最初，搜救行动由吉隆坡航空救援协调中心负责协调，2014 年 3 月 8 日至 15 日在马来半岛以东的中国南海进行。在此区域进行搜救的主要原因是这里就是与客机最后联络及二次雷达应答机停止运行的地方。

马来半岛西部的情况

7. 2014 年 3 月 8 日至 15 日,搜寻行动在马来半岛以东的中国南海进行,同一时间由马来西亚皇家空军协调的搜救行动也在马来半岛西部进行。

8. 2014 年 3 月 8 日至 15 日,搜寻行动也在马六甲海峡、安达曼海、孟加拉湾和苏门答腊西部展开。搜寻此区域的原因是原始雷达数据显示当时有一架国籍不明的飞机飞越马六甲海峡,大家认为这架飞机就是马航 370 客机。

发现北南两条空中走廊

9. 在海面搜救的第一周,国际海事卫星组织卫星通信对马航 370 的数据分析表明,客机最后一次在苏门答腊北端被雷达捕获后又飞行了 6 小时。最初的卫星数据分析表明,客机当时是沿两条空中走廊中的一条飞行:一条朝北,向哈萨克斯坦方向,另一条朝南,向印度洋方向。掌握这个新信息后,马来半岛西部和东部的搜救行动于 2014 年 3 月 15 日暂停。

10. 2014 年 3 月 18 日至 23 日的卫星通信数据表明,吉隆坡航空救援协调中心、巴丹国家搜寻救援局以及印度尼西亚共和国国家搜寻与救援局在南部空中走廊区域内进行了空中搜寻。

澳大利亚主导的海面搜寻

11. 2014 年 3 月 17 日,由于南部走廊延伸至印度洋和澳大利亚境内的搜救区域,应马来西亚的要求,澳大利亚方面承担了这次搜救行动的协调工作。澳大利亚海事安全局负责协调搜救行动,部署飞机和水面船舶从西澳大利亚开始行动。本次行动于 2014 年 4 月 28 日结束。在搜寻区域方面,澳大利亚海事安全局继续听取联合调查组和卫星通信工作组的建议。这些建议是对国际海事卫星通信日志和此机型其他客机的性能分析逐步分析得出的。

12. 此次海面搜寻的重点是识别和打捞任何浮在海面的马航 370 客机残骸。澳大利亚海事安全局接管海面搜寻的协调工作时,客机已经失踪了九天。因此,有必要通过分析来得出客机可能中止飞行的位置、计算自 2014 年 3 月 8 日后的几日内客机各种类型残骸漂移的范围,来确定飞机和水面船舶搜寻的区域。

13. 澳大利亚海事安全局为此成立了一个漂移模拟工作组,以确保为搜救行动提供最佳漂移模拟。工作组由澳大利亚联邦科学与工业研究组织、亚太应用科学协会、美国海岸警卫队和气象与全球环境模拟与监测系统局等多个组织组成。在海面搜寻行动的全过程中,由飞机和船只部署的自定位基准标志浮标为漂移模拟提供信息。自定位基准标志浮标是一种海面漂流浮标,装有用于测量表层洋流的全球定位系统。部署在海洋上的标志性浮标预计可使用 20 多天。同样,来自搜寻区域的实时风浪数据也不断为漂移模型更新信息。

【注释】

① 本句中介词短语 "in relation to the aircraft's most likely flight path" 作后置

定语,修饰"progressively refined analysis",翻译时要注意这个比较长的短语各部分之间的关系,译成"依照该飞机最有可能的飞行路线随时细化地分析"。

② "Malaysian led surface search"从结构上讲是一个主谓宾结构,在文中充当标题,最好处理成汉语名词短语的形式:马来西亚主导的海面搜寻。类似的还有Australian led surface search(澳大利亚主导的海面搜寻)。

③ 本句结构比较复杂,句子主干是一个被动结构 This area was searched…,其后 based on the primary radar data which indicated an unidentified aircraft had flown up the Strait of Malacca, thought to be MH370 作状语,状语中 which 引导的定语从句修饰 primary radar data,非限制性结构 thought to be MH370 对 an unidentified aircraft 做进一步的说明。翻译时,先要缕清句子结构,增加"当时""大家"等词,使汉语的表述更有层次,条理更清晰。

④ 本句中"with"引导的含现在分词的独立结构"with the southern corridor extending into …"在句中作原因状语,翻译成"由于……",将原文隐性的逻辑关系显化出来。

⑤ 本句"using aircraft and surface vessels operating from Western Australia until 28 April 2014"部分,using aircraft and surface vessels 是非谓语动词短语作状语,表示方式,对主句部分进行进一步描述。"operating from …"作后置定语,修饰"aircraft"和"vessels"。

⑥ 此句使用了无人称句"It was (therefore) necessary to…"。这种句式在科技英语中比较常见,目的是突出科技文体的准确性和客观性。科技英语中常见的无人称句除了这种"It…that(to)…"的句式外,还有以下两种:

a. 带有表语从句的倒装句,如 The fact/case/question/is that…、The chances are that…、Of importance/concern is that…、Particularly noteworthy is that…

b. 带有宾语从句的名词化结构,如 Study shows that…、Results demonstrate that…、Tests have proved that…等。本句结构非常复杂,一环套一环:理解时,需要仔细辨析,分清各部分之间的关系;翻译时,可先将长句按意群断开,然后按照汉语表达习惯重新排列,再通过添加连接词"因此"等,使汉语的表达通顺流畅。

⑦ 本文出现了大量的专有名词如国名、地名、机构名称以及科技术语的缩略语,如本句中的 CSIRO、Asia-Pacific Applied Science Associates、the United States Coastguard、the Bureau of Meteorology and Global Environmental Modelling and Monitoring Systems 等。翻译时务必勤查字典和资料,采用已经约定俗成的汉语译法,切不可望文生义,随意翻译。

综合实践4 航空航天科技新闻报道

科技新闻报道一般由标题、导语和正文组成。英语新闻报道的标题具有简明、新

奇的特征,内容简短,重点突出、往往省略系动词、助动词、冠词等。例如:

Airline studies Jet Fuel Contamination

航空公司正在研究喷气燃料受污染问题

Planning a Planetary Defence Against Asteroids

抵御小行星,打响地球保卫战

China Announces Successful Launch of 2nd Lunar Probe

中国宣布成功发射第二颗月球探测器

科技新闻报道的受众为一般读者,主要特点为用词通俗易懂,语句简练明确,报道客观翔实,一般不涉及深奥的专业理论。少数情况下会不可避免地使用专业词汇,但大多数是该专业的基础词汇。译文要力求再现原文的风格,并与原文在格式上保持一致,做到表意明确,结构完整,通俗易懂。

翻译练习 4.1

Inspiration4: All-Amateur Space Crew[①] Enjoy Views of Earth

1. The four private fliers are pictured floating around inside their space capsule and looking out the vehicle's big domed window at the planet below.

2. Billionaire Jared Isaacman and his colleagues Hayley Arceneaux, Sian Proctor and Chris Sembroski launched from Florida on Wednesday. [②]

3. They're the first all-amateur crew to go into orbit on a commercial mission.

4. Their trip was purchased from, and is organised by, tech entrepreneur Elon Musk's SpaceX rocket company. [③]

5. The California-based firm used one of its Dragon ships to lift the foursome to an altitude as high as 590 km. That's about 160 km further than the International Space Station, and 50 km higher even than the famous Hubble Space Telescope.

6. Not much information has been released since launch about the activities of the crew, who are the subject of an exclusive Netflix documentary. [④]

7. It's known that[⑤] they are well, having adjusted to their weightless environment, and have been conducting some scientific experiments.

8. It's also known that they have been in conversation with family and friends on Earth, and have conducted a Q&A with patients at St Jude Children's Research Hospital in Memphis, Tennessee.

9. Ms Arceneauxis a physician assistant (known as a physician associate in the UK) at the hospital, having also been treated there as a 10-year-old for bone cancer.

10. Mr Isaacman hopes the interest around the Inspiration 4 flight can help raise $ 200 m for St Jude.

11. The crew are due back on Earth at the weekend. Their capsule will bring them to a splashdown in the Atlantic Ocean.

12. It will have been another milestone in the space tourism market, which is experiencing a resurgence after a decade's hiatus.

13. Earlier this summer, billionaire businessmen Sir Richard Branson and Jeff Bezos went above Earth's atmosphere in their own space vehicles.

14. The coming months will see similar trips.

【参考译文】

灵感4号："全平民"太空旅行团欣赏地球景色

1. 摄像机捕捉到了四名平民航天员在太空舱内四处飘荡,透过玻璃穹顶遥望下方的行星。

2. 周三,亿万富翁贾里德·艾萨克曼和他的同伴黑莉·阿尔塞诺、希恩·普罗科特和克里斯·森布罗斯基所搭乘的飞船在佛罗里达州发射升空。

3. 他们是首批通过商业太空载人飞行任务进入轨道的全平民机组人员。

4. 此次太空旅行是从科技企业家埃隆·马斯克的美国太空探索技术公司购得,并由该公司负责组织。

5. 总部位于加利福尼亚的美国太空探索技术公司使用龙飞船将四人旅行团送到距地面约590千米高空,比国际空间站高出近160千米,甚至比著名的哈勃太空望远镜还要高50千米。

6. 网飞公司独家纪录片负责记录全平民太空旅行团的活动,但自飞船发射以来,并未发布多少有关他们的活动信息。

7. 据悉,他们状态不错,已经适应了失重的环境,并一直在做科学实验。

8. 据了解,他们已经和地球的家人和朋友进行了交谈,还与田纳西州孟菲斯圣裘德儿童研究医院的病患进行了问答交流互动。

9. 阿尔塞诺女士是这家医院的医师助理(即英国的医生助理),她10岁时曾因骨癌在那里接受治疗。

10. 艾萨克曼先生希望公众对灵感4号飞行任务所产生的兴趣,能为圣裘德儿童研究医院筹款2亿美元。

11. 预计周末太空旅行团将返回地球。他们乘坐的太空舱将降落大西洋洋面。

12. 在停滞十年后,太空旅游市场又开始复苏,而这次旅行将是一个新的里程碑。

13. 今年夏天早些时候,亿万富翁理查德·布兰森爵士和杰夫·贝索斯也曾搭乘各自的太空飞船飞越了地球大气层。

14. 未来几个月,还会有类似的太空商业旅行。

【注释】

① "amateur"原意"业余的,非专业的",通览全文后可以看出此处的"all-amateur space crew"是指平民太空游客。"crew"原意"机组人员",放在此处可以引申为"游客团",既将其深层意义翻译出来,又生动形象,作为标题引人注目。

② 此句的主语为人,而动词直接用了"launch",取"发射,起飞"一义时动作主体常为火箭、卫星等物,所以翻译时需补充"……所搭乘的飞船在佛罗里达州发射升空"。

③ 此处的两个动词"be purchased from"和"be organized by"虽表面主语均为SpaceX,但翻译时需注意前者主语实为四名太空游客,因此需分开翻译。

④ 原句主语为"Not much information",但翻译时需重新选定主语。而选用"Netflix documentary"为主语可以顺理成章地将被动语态转为主动语态。

⑤ "It is known that"一般译为"众所周知",而此处译为"据悉",更加符合本文新闻报道的文风,与下段的"据了解"相对应。

翻译练习 4.2

NASA Makes Blue Origin Eligible to Launch Future Missions Without Crews

1. Jeff Bezos and his company's New Glenn rocket, which hasn't yet flown, get nod to potentially carry scientific payloads① later this decade.②

2. The space transportation company run by Amazon. com Inc. Chief Executive Jeff Bezos reached another important steppingstone③ in its effort to become an established launcher of U. S. civilian and national-security payloads.

3. The National Aeronautics and Space Administration on Wednesday named Blue Origin Federation LLC's New Glenn rocket—a 310-foot reusable booster—as one of its potential launch providers for scientific missions later in the decade.④

4. The agency said no specific contracts had been awarded to Blue Origin, but the announcement for the first time makes the closely held company eligible for such NASA business. Under the arrangement, the company Mr. Bezos founded nearly two decades ago will be allowed to use New Glenn, roughly six stories taller than rival rockets, to compete for awards. Various NASA centers will be able to design spacecraft to take advantage of New Glenn's power and other features, including its capability to transport larger payload volumes than other rockets.

5. In a statement, Blue Origin said:"We are proud to be in NASA's launch services catalog and look forward to providing reliable launches" for NASA for

years to come.⑤

6. Other heavy-lift rockets already cleared to compete for NASA's scientific launches include those operated by Elon Musk's Space Exploration Technologies Corp. and a joint venture between Boeing Co. and Lockheed Martin Corp.⑥

7. New Glenn，which is receiving development funding from the Pentagon，also is in the runningfor national-security launches. And Blue Origin previously signed up a number of commercial-satellite operators as customers.

8. Until the past few years，Mr. Bezos maintained a shroud of secrecy around Blue Origin by avoiding publicity, personally investing roughly ＄1 billion in the company in some years while declining to discuss test launches of its smaller New Shepard booster before its first blast off in 2015.

9. In addition to transporting satellites，Blue Origin has been testing hardware to take space tourists on suborbital thrill rides.⑦

10. But as the company grew and the Boeing-Lockheed joint venture agreed to buy Blue Origin engines to help power its next-generation booster targeting Pentagon contracts⑧，Mr. Bezos and his team opened up about their plans.

11. In a 2019 interview，Chief Executive Bob Smith said the company's strategy relied heavily on winning lucrative military and other government business for its mega-rocket. "We need those customers," he said.

12. Since then，Blue Origin has successfully bid to become one of three teams devising lunar landers intended to transport NASA astronauts⑨.

13. Blue Origin has constructed a sprawling assembly facility at Cape Canaveral，Fla.，where it plans to build, service and launch a fleet of New Glenn boosters. Representing what people in the industry have estimated amounts to a ＄2.5 billion investment，New Glenn is designed to lift up to 45 tons into low-earth orbit—a third more tonnage than SpaceX's largest rocket currently, the Falcon Heavy.

【参考译文】

美国国家航空航天局准予蓝色起源公司未来承担非载人火箭发射任务

1. 杰夫·贝索斯及其公司尚未执行任何任务的"新格伦号"火箭获得美国国家航空航天局许可,有望在未来三五年运载科学载荷进入太空。

2. 这家太空运输公司由亚马逊公司首席执行官杰夫·贝索斯经营,致力于成为一家认定的火箭发射商,负责承运美国民用及国家安全设备进入太空。美国国家航空航天局的这一许可意味着该公司进入了一个新的发展阶段。

3. 周三,美国国家航空航天局将蓝色起源联盟有限责任公司的"新格伦号"火箭指定为未来三五年可能承担发射任务的运载火箭之一。该火箭长 310 英尺,可重复使用。

4. 美国国家航空航天局表示他们还未与蓝色起源签订具体合同,但公告一出,即说明这家控股公司首次拥有了承接美国国家航空航天局此类业务的资格。按照安排,这家近 20 年前由贝索斯先生创立的公司就可以使用"新格伦号"火箭参加竞标。该火箭差不多比竞争对手高出 6 层楼。同时,美国国家航空航天局各研发中心也能在设计航天器时充分借鉴"新格伦号"的动力和其他各种特性,包括优于其他火箭的更大有效载荷运载量。

5. 在一份声明中,蓝色起源表示能名列美国国家航空航天局发射服务目录,他们感到非常自豪并期待在未来几年为美国国家航空航天局提供可靠的发射服务。

6. 埃隆·马斯克的太空探索技术公司以及波音公司和洛克希德·马丁公司的一家合资企业旗下的重型火箭也获准参与竞争,争取拿下运送美国国家航空航天局科学设备的资格。

7. 已获得美国国防部开发基金的"新格伦号",同时也参与角逐美国国家安全设备的发射任务。在此之前,蓝色起源已经与一些商业卫星运营商成功签约。

8. 直到近几年,贝索斯先生还刻意让蓝色起源保持一定的神秘色彩:他不搞宣传推广;他先后个人投资蓝色起源约 10 亿美元;对测试发射小型化的"新谢泼德"助推火箭,他也闭口不谈,直到这个火箭 2015 年正式发射升空。

9. 除了发射卫星外,蓝色起源还不断测试火箭硬件,希望有朝一日能将游客送入太空,让他们体验惊险刺激的亚轨道之旅。

10. 为了获得美国国防部的合同,波音-洛克希德合资公司同意购买蓝色起源的发动机,为其新一代助推火箭提供动力,再加上蓝色起源自身的发展壮大,贝索斯及其团队终于将公司的各项计划公之于众。

11. 在 2019 年的一次采访中,蓝色起源首席执行官鲍勃·史密斯表示该公司的经营方针就是为其巨型火箭尽可能地争取利润丰厚的军方和政府业务。"我们需要那些客户,"他说。

12. 至今,蓝色起源公司已经成功跻身月球飞行器研制三大团队。这些月球着陆器未来会运载美国国家航空航天局航天员飞往月球。

13. 蓝色起源公司已经在佛罗里达州卡纳维拉尔角建造了一个庞大的火箭设备组装基地,并计划在那里建造、维修和发射一系列"新格伦号"助推火箭。据业内人士估计,总投资达 25 亿美元的"新格伦号"火箭,其设计目标是能将重达 45 吨的有效载荷送入近地轨道——这比美国太空探索技术公司目前最大的重型火箭"猎鹰号"的运载能力还多三分之一。

【注释】

① "payloads"原指飞行器等运输工具的收费载重,后指有效载荷,即航天器上装

载的为直接实现航天器在轨运行要完成的特定任务的仪器、装置、人员、试验生物及试件等。此处将"scientific payloads"翻译为"科学载荷",简洁明了,对读者来说清楚易懂。

② "later this decade"中"decade"意为"十年",但此处表达为虚指,意义相当于汉语中的约数,因此处理为"三五年"。

③ "reached another important steppingstone"中"steppingstone"一词的字面意义为"踏脚石、进身之阶",结合句意,实指经历了这一重要发展阶段,因此将其处理为"进入了一个新的发展阶段"。

④ 长句处理。先找出句子主干"The National Aeronautics and Space Administration on Wednesday named Blue Origin Federation LLC's New Glenn rocket…as one of its potential launch providers for scientific missions later in the decade"及句中的插入语,本着汉语"事实在前,解释在后"的原则,先翻译主干,后翻译插入语并独立成句。

⑤ 原文的直接引语与间接引语格式重叠,稳妥起见翻译时宜统一处理为间接引语,确保表意清晰同时避免争议。

⑥ 此句可按照原文语序翻译,但未免定语修饰的主语过长,宜调整主语为各公司的重型火箭,同时补充翻译"争取拿下运送美国国家航空航天局科学设备的资格",使句意连贯。

⑦ 原文句子较短,但翻译时需把握句意,将目的状语单独处理,并添加"送入""体验"等动词,将原文的名词短语动词化,使得译文更加生动。

⑧ 语序调整。将"and"并列连接的较长的部分提前翻译,使得句意通顺连贯。

⑨ 合理断句。若按原文语序直译,那么"lunar lander"的修饰语过长,不经断句难免繁复啰嗦,因此修饰语单独成句翻译。

翻译练习 4.3

Watch① SpaceX's Starship Launch and Explode in Crash Landing

1. The company described the test of the next-generation spacecraft as "awesome" even though it ended in a fiery blast.

2. A test of a prototype of a rocket that Elon Musk has dreams of sending people to Mars in flew several miles high on Wednesday.② But in attempting to land, it hit the ground too fast and exploded.

3. That was the latest test, partly successful, in the development of next-generation spacecraft built by SpaceX, Mr. Musk's rocket company.③

4. SpaceX's live broadcast showed the smoldering remains of the rocket, named Starship, at the company's test site in southern Texas. "Awesome test,"

read text across the screen of the broadcast after the fiery conclusion. "Congratulations, Starship team!"④

5. With the late afternoon sun low on the horizon, the gargantuan stainless steel spacecraft lifted off from a launch pad at 5:45 p. m. After reaching its apogee, it started falling, as planned, tipping over in a controlled glide back to Earth.⑤

6. Near the ground, it righted itself back to a vertical orientation and fired its three engines to slow down—but not enough. Upon impact, about 6 minutes, 40 seconds after liftoff, it disintegrated in a fireball, leaving a cloud of smoke rising over the test site.

7. Still it was an advance in SpaceX's development efforts. Earlier prototypes disintegrated without ever leaving the ground during tests that simply pumped ultracold liquid propellants into the rocket's tanks.⑥

8. SpaceX has become successful in the launch business, and it is now the world's most valuable privately held company. Its Falcon 9 rockets have become a dominant workhorse⑦ for sending satellites to orbit. It routinely transports cargo to the International Space Station, and has lifted NASA astronauts there twice this year, with more trips planned in 2021.

9. However, many are skeptical when Mr. Musk says the company is just a few years away from sending a Starship to Mars, and he has repeatedly set timelines for SpaceX that proved far too optimistic in how quickly they have come to pass.

10. Last year, when he provided an update on the development of Starship, he said a high-altitude test would occur within months and orbital flights could occur early this year.

11. Instead, several catastrophic⑧ failures happened because of faulty welding. When the propellant tanks stopped rupturing, one of the prototypes made a short successful flight in September. That earlier Starship model, which resembled a spray paint can with the label removed, lifted itself using a single rocket engine nearly 500 feet before setting down at the Texas test site near the village of Boca Chica.

【参考译文】

直击美国太空探索技术公司星际飞船发射及紧急着陆爆炸现场

1. 尽管新一代航天器最后发生猛烈爆炸,美国太空探索技术公司仍认为这场测试 "可圈可点"。

2. 埃隆·马斯克一直梦想着用火箭将人类送往火星。周三的飞行测试中,这种火箭的原型机在升空数英里后,却因着陆时速度过快发生爆炸。

3. 虽然功败垂成,这却是马斯克创办的美国太空探索技术公司研发新一代宇宙飞船进程中最新的一次尝试。

4. 美国太空探索技术公司的现场直播显示,在得克萨斯州南部的公司测试场上,"星舰号"火箭的残骸仍浓烟滚滚。测试以爆炸告终后,直播节目屏幕上出现一行字:"令人惊叹的测试,祝贺星舰号研究团队!"

5. 傍晚5点45分,夕阳西下。这艘巨型不锈钢飞船发射升空,在到达远地点后,开始按计划降落:飞船折返,平稳滑行,返回地球。

6. 在接近地面时,星舰号重新调整成垂直姿态,并启动了其三台发动机来减速——但并未奏效。在发射约6分40秒后,火箭猛烈撞击地面,机体碎片横飞化作一团火球,测试场上烟雾升腾。

7. 尽管本次试飞未能成功,但仍是美国太空探索技术公司火箭开发领域的一次进步。在以往的测试中,原型机都未曾离开过地面,在测试进行到将超冷液体推进剂注入火箭燃料舱时,就发生解体了。

8. 美国太空探索技术公司在火箭发射行业取得巨大成功,如今已成为全球最有价值的私人控股公司。该公司的"猎鹰9号"火箭已成为将卫星送入轨道的主要运输工具。这种火箭定期为国际空间站运输货物,今年已经两次将美国国家航空航天局的航天员送至空间站,并计划在2021年完成更多运输任务。

9. 马斯克先生称过不了几年,公司就能向火星发射"星舰号"飞船,但许多人对此将信将疑。他一再为美国太空探索技术公司制定时间表,但事实证明,他对这些时间表的实施进度过于乐观。

10. 去年,马斯克在发布有关星舰号开发的最新进展时说,几个月内将会做一次高空测试,今年早些时候也可能会进行多次轨道飞行。

11. 但由于焊接问题,几次测试都失败了,损失惨重。9月,燃料舱破裂问题解决后,一架原型机终于成功进行了短途飞行。这个早期的星舰样机看上去就像一个去掉了标签的喷漆罐,依靠单个发动机飞到了近500英尺(152.4米)高度,然后降落在博卡奇卡村附近的得克萨斯试验场。

【注释】

① "Watch"译为"直击"更具现场感和代入感,使文章标题更有特色,吸引读者。

② 断句处理。原句的定语修饰词过长,宜作为背景单独成句翻译。

③ 将陈述句式译为汉语的转折关系句,灵活处理"test"的后置定语"partly successful"。这样,逻辑明晰,更具可读性。

④ 原文的两处直接引语一前一后,而译文将其放在一句中,更符合汉语的用语习惯。

⑤ 此处译文用":"表明对下文的概括包含关系。

⑥ 此处长句翻译时需先厘清动作发生的先后顺序，并恰当体现原文语气，先翻译事件过程，将"disintegrated"这一结论放在句末。

⑦ "workhorse"：指"Falcon 9 rockets"，根据文意翻译为"运输工具"。

⑧ 将形容词"catastrophic"单独翻译，放在文末，更符合汉语表达习惯。

翻译练习 4.4

Satellites Spot Oceans Aglow with Trillions of Organisms

1. A new generation of detectors let scientists identify a dozen large episodes of bioluminescence, one a hundred times larger than Manhattan—and that's the smallest.

2. The ocean has always glowed.

3. The Greeks and Romans knew of luminous sea creatures as well as the more general phenomenon of seawater that can light up in bluish-green colors①.

4. Charles Darwin, as he sailed near South America on a dark night aboard the H. M. S. Beagle, encountered luminescent② waves③. He called it "a wonderful and most beautiful spectacle." As far as the eye could see, he added, "the crest of every wave was bright"—so much so that the "livid flames" lit the sky.

5. Now, scientists report that ocean bioluminescence can be so intense and massive in scale that satellites orbiting five hundred miles high can see glowing mats of microorganisms as they materialize in the seas. Last month in the journal Scientific Reports, eight investigators told of finding a luminous patch south of Java in 2019 that grew to be larger than the combined areas of Vermont, New Hampshire, Massachusetts, Rhode Island and Connecticut.

6. "It was an epiphany④," said Steven D. Miller, lead author on the bioluminescence study and a specialist in satellite observations at Colorado State University. When a hidden wonder of nature comes to light, he added, "it captures your imagination."

7. The scientists said the close examination of images gathered between December 2012 and March 2021 from a pair of satellites let them identify a dozen extremely large events—approximately one every eight months. Even the smallest was a hundred times larger than Manhattan.

8. The imagery is opening a new window⑤ on the world's oceans, scientists say, and promises to aid the tracking and study of the glowing seas, whose origins are poorly understood.

9. Kenneth H. Nealson, a pioneer of bioluminescence research at the

University of Southern California, called the discovery"a big step toward being able to understand" how an enduring mystery of the sea "actually comes to be."⑥

10. The new paper noted that the large concentrations of living lights have long "eluded rigorous scientific inquiry, and thus little is known about their composition, formation mechanism, and role within the marine ecosystem."

11. Sea bioluminescence is often associated with gruesome creatures of the inky deep. One iconic illuminator is the anglerfish, which dangles bright lures⑦ in front of needlelike teeth. In contrast, the luminous seas appear to originate when many trillions of tiny bacteria light up in unison.

12. Dr. Nealson, who was not involved in the satellite research, and colleagues reported in 1970 that dilute suspensions of a particular type of bacteria emit no gleam. If allowed to multiply, however, the microbes can suddenly light up as if a switch were thrown. Scientists now theorize that the glittering masses of bacteria lure fish, whose guts provide nourishing habitats.

13. Dr. Miller's trail of discovery began nearly two decades ago when a lunchtime chat raised the question of whether sea bioluminescence might be visible from space⑧. While working at the U. S. Naval Research Laboratory in Monterey, Calif., in 2004, he began examining imagery from a weather satellite. Soon, he spotted in the northwestern Indian Ocean what turned out to be a glowing patch that was nearly the size of Connecticut.

14. The blurry area was barely visible, but Dr. Miller and his colleagues grew quite excited because they knew that a new generation of satellite sensors would soon provide much greater sensitivity and sharpness⑨. The improved sensors debuted on a pair of satellites launched by the National Oceanic and Atmospheric Administration in 2011 and 2017.

15. The sensitive detectors proved adept—at least on dark nights—at capturing glimmers of light from the seas and provided the imagery for the current report.

16. A surprise, Dr. Miller said, is that the events turn out to persist for long periods. For instance, the large patch off Java in 2019 lasted for at least 45 nights. That raises the possibility that a rapid response team of oceanographers might have enough time to reach the patches and take samples for detailed studies.

17. To date, Dr. Miller said, no team has succeeded. He added that television companies that make nature documentaries have shown interest in using the satellite detections to track down and film the gleaming seas.

18. Peter Herring, a British marine biologist known for his work on deep bioluminescence, calledthe satellite work important because, after ages of

uncertainty⑩, it raised the prospect of finally coming up with hard evidence of what powers the luminous whorls.

19. The discovery, he added, "is a large splash and will have significant ripples."

【参考译文】

卫星发现数万亿海洋发光生物

1. 科学家使用新一代探测器发现了 12 起大型生物发光事件,其中最不起眼的一起事件覆盖面积也比整个曼哈顿大百倍。

2. 海洋一直在发光。

3. 古希腊人和古罗马人对发光海洋生物已有所了解,同时他们也懂得海水凸显绿色偏蓝光彩这一更常见的现象。

4. 在一个漆黑的夜晚,查尔斯·达尔文搭乘皇家海军舰艇"贝格尔号"在南美洲附近航行时,意外地发现了海浪发光。他称之为"绝妙至美景象"。他还说,在目之所及的地方,"每一个浪尖都闪亮着"——汇聚起的"青灰色火焰"点亮夜空。

5. 目前,科学家报告称,海洋生物发光事件的强度和规模大到距地面 500 英里(800 千米)高度轨道上运行的卫星都可以捕捉到微生物在海洋中呈现出的发光团。据《科学报告》杂志上月报道,8 名调查人员讲述了 2019 年在爪哇岛以南发现的发光区域,后来其面积增大,超过佛蒙特州、新罕布什尔州、马萨诸塞州、罗德岛和康涅狄格州的总和。

6. 生物发光研究的首席作者、科罗拉多州立大学卫星观测专家史蒂文·D·米勒说,"这如同神灵显现一般"。他说,隐秘的自然奇观一旦出现,"便会激发你的想象力"。

7. 科学家们透露,通过对 2012 年 12 月至 2021 年 3 月之间一对卫星收集的图像进行仔细研究,他们认定了 12 起特大生物发光事件——大约每 8 个月就发生一起。其中最小的一起事件,其面积也是曼哈顿的百倍。

8. 科学家们指出,这种发光现象为了解全球海洋开辟了新的途径,有望帮助我们对其起因知之甚少的海洋发光事件进行追踪和研究。

9. 南加州大学生物发光研究的开拓者肯尼斯·H·尼尔森将这一发现称之为在"朝着能够理解"海洋永久之谜"如何真正形成"方面"迈出的一大步"。

10. 这篇新发表的论文还指出,生物发光的大规模聚集现象长久以来一直没有经过严密的科学研究,因此我们对其构成、形成机制和在海洋生态系统中的作用知之甚少。

11. 海洋生物发光通常与漆黑深海中的可怕生物联系在一起。华脐鱼就是一种标志性的发光体,它针状牙齿前悬荡着用来捕获其他鱼类的明亮诱饵。倒是数万亿极小细菌同时发光时,海洋发光现象便产生了。

12. 并未参与卫星研究的尼尔森博士和他的同事在 1970 年报告称,某种特殊类型细菌的稀释悬浮液本身不会发光。然而,如果听任这种微生物大量繁殖,它们就会像打开了开关一样突然明亮起来。科学家现在提出理论:发光的菌团引诱鱼类捕食,而鱼类的肠道又为菌群提供了养分充足的栖息地。

13. 米勒博士的发现之旅始于大约 20 年前的午餐闲聊,交谈间海洋生物发光现象在太空中是否可见这个问题浮出水面。2004 年在加州蒙特利美国海军研究实验室工作期间,他着手研究气象卫星图。不久,便在印度洋西北部发现了一片确实正在发光的区域,面积接近康涅狄格州。

14. 尽管那片模糊的区域勉强可见,但米勒博士和同事们却非常兴奋,因为他们知道很快新一代的卫星传感器会提供远高于以往的灵敏度和锐度。在 2011 年和 2017 年,改进后的传感器首次在美国国家海洋和大气管理局发射的两颗卫星上亮相。

15. 至少在漆黑的夜晚,灵敏的探测器的确能敏捷地捕捉海洋发出的缕缕微光,并为当前的报告提供了图像。

16. 米勒博士说,令人惊讶的是这些现象最终会持续很长时间。例如,2019 年爪哇岛上的一大片区域至少连续 45 个夜晚发光。这样,海洋学家快速反应小分队更有机会获得充足的时间到达发光区域,采集样本进行深入研究。

17. 米勒博士说,到目前为止,还没有一个小分队成功。他说,摄制自然类纪录片的一些电视公司表达了对利用卫星探测来追踪和拍摄发光海域的兴趣。

18. 研究深海生物发光的英国海洋生物学家彼得·赫林认为卫星的运用很重要,因为经过漫长的迷茫岁月,卫星的运用最终让人类有望最终找到海洋菌群缘何发光的有力证据。

19. 他说,这一发现将会引起巨大轰动,影响深远。

【注释】

① "bluish-green colors"译为"绿色偏蓝光彩"。在进行颜色词翻译时,应灵活处理,避免错译或死译,注意以下三个方面:

a. 一些汉语颜色词可能对应多个英语颜色词,因此应结合上下文内容确认对应的表达。如:blue sky 碧空;white cloud 碧云;green jade 碧玉。

b. 由于民族文化差异,英汉中可能使用不同词表述同一种事物的颜色。例如:black tea 红茶;grey hair 白发;black and blue 又青又紫。

c. 有时为了更加准确地突出原文含义,可采用"省其色,译其意"的方式,可以直接翻译其深层内涵。

如:the fish too green 鲜龙活跳;born in the purple 门第显赫。

② 此处根据语境,将形容词"luminescent"译为动词"发光",更符合汉语表达习惯。

③ 此句比较长,需重新划分结构并调整语序,故将时间状语"on a dark night"提前,汉语读起来更通顺。

④ "epiphany"是基督教用语,耶稣显灵的意思。此处作者用来表达对海洋发光现象的惊叹。

⑤ "opening a new window"若直译为"开辟新窗",过于生硬,不符合汉语表达习惯,应联系上下文,译为"开辟了新的途径"。

⑥ 此处句式较为复杂,应厘清句子主干"called the discovery a big step",然后确定各部分的修饰关系,层层进行翻译。

⑦ 此处应补充翻译"用来捕获其他鱼类的",帮助读者理解"lure"的具体用途。

⑧ 该句较长,应重新划分结构,将"raised the question"及后面的从句分成小短句进行解释。

⑨ "sensitivity and sharpness"指"灵敏度和锐度(或对比度)"。

⑩ Uncertainty 本意为不确定、难以预料,这里指茫然、无从寻求答案的状态。"ages of uncertainty"译为"经过漫长的迷茫岁月",表达生动,更有意境。

翻译练习 4.5

China Releases Rover's First Photos After Mars Landing①

1. The country's space agency said that the components of the spacecraft had "deployed in place normally."

2. Four days after landing a spacecraft on Mars,② China's space agency released its first photographs from the red planet on Wednesday, announcing that the mission was going as planned.

3. The four-day wait for the images—one in color, one in black and white, as well as a pair of small video clips③—had prompted speculation④ that something might have gone wrong with the landing on Saturday. China's space agency issued a statement in response to those concerns on Tuesday, urging patience.

4. After decades of exploring Mars, NASA has a flotilla of spacecraft in orbit around the planet to relay data from its rovers, Perseverance and Curiosity, that are driving on the surface. The Chinese possess no existing spacecraft infrastructure at Mars to take advantage of.

5. The Tianwen-1 mission consists of an orbiter,⑤ lander⑥ and rover that launched in July and arrived at Mars in February. ⑦ But the orbit that put Tianwen-1 in position to release the lander and rover last week was not ideal for relaying large chunks of data, like images and videos, back to Earth.

6. On Monday, the orbiter fired its thrusters, and it now circles Mars once

every eight hours instead of two days. That allows more frequent and faster communications with the rover, which has been named Zhurong, after a mythical Chinese god of fire.⑧

7. The photographs were the first public evidence that China's lander had successfully reached the surface.

8. The landing made China only the third nation to touch down safely on Mars, after the United States and, very briefly,⑨ the Soviet Union. It was the latest in a series of major milestones, including missions to the moon and the start of construction of a new orbiting space station, that have secured China's status as a space power. (China may also launch a second module for the space station this week.)

9. Since the landing craft reached the surface on Saturday, the China National Space Administration had divulged little about the Mars mission's progress. It⑩ said on Wednesday that the components of the lander and the rover, including its solar panels, had "deployed in place normally. "

10. The black-and-white photograph shows the ramp that will guide the rover off the lander to the surface, casting a crisp shadow on the surface. (The horizon's arc is an effect of the wide-angle lens.)

11. The other image, in color, shows the rear of the rover and signs that the solar panels that will power the vehicle had unfolded successfully. In the background are the red rocks and soil of Utopia Planitia,⑪ the impact basin where NASA's Viking 2 probe also landed in 1976.

12. "The landscape is totally different than that of the Viking Lander 2 site, on the northern side of the Utopia Basin,⑫ which has lots of boulders and rocks," said James W. Head III, a professor of geological sciences at Brown University.

13. "The terrain around Zhurong looks flatter and boulderless⑬ with some interesting rocks in the foreground but mostly fine-grained sediment. " Dr. Head said.

14. "This could be sediment emplaced in an ancient Mars ocean, and it could be glacial and ice-related materials too," Dr. Head said.

15. The Chinese agency also released two brief videos of the lander departing the orbiter that carried the craft to Mars. ⑭

16. Zhurong is carrying a number of instruments to study the planet's topography, geology and atmosphere. One goal is to understand the distribution of ice in the region, which, in theory, could someday help sustain visits by people. It is expected to drive off the lander in a few days.

17. In a statement released by NASA on Wednesday, Bill Nelson, the agency's administrator, congratulated the Chinese space agency on the first images.

18. "As the international scientific community of robotic explorers on Mars grows, the United States and the world look forward to the discoveries Zhurong will make to advance humanity's knowledge of the Red Planet," Mr. Nelson said in the statement.

【参考译文】

中国发布火星车着陆后首批照片

1. 中国国家航天局表示，火星探测器的各部件已"正常部署到位"。

2. 探测器着陆火星四天后，国家航天局周三发布了来自这个红色行星的首批照片，宣布探测任务正按预定计划进行。

3. 一张彩色照片、一张黑白照片以及两段短视频：等待了四天才等来的这些照片和视频，不免让人们猜疑周六的着陆可能出了问题。国家航天局周二发表声明，对这些担忧作出回应，呼吁人们保持耐心。

4. 在对火星进行了几十年的探索后，美国国家航空航天局已在火星轨道上有一个探测器小型编队，能为在火星表面行驶的"毅力号"和"好奇号"火星车做数据中继传输。而中国在火星没有现存的航天器基础设施可利用。

5. "天问一号"探测器由环绕器、着陆器和火星车组成，于去年7月发射，并于今年2月抵达火星。但"天问一号"上周分离着陆器和火星车所在的轨道，并不特别适合将图像和视频等大量数据传回地球。

6. 周一，环绕器启动了推进器，现在它八小时环绕火星一周，而之前则需两天。这样，环绕器便可以与火星车进行更频繁、更高速的通信。火星车以中国神话中的火神"祝融"命名。

7. 这些图像是中国着陆器成功降落火星表面的首批公开证据。

8. 这次着陆让中国成为继美国和苏联之后，第三个在火星上安全着陆的国家，苏联曾非常短暂地着陆于火星。这次成功着陆火星，连同探月任务和开始建造轨道空间站，成为一系列重大里程碑事件，巩固了中国太空强国地位。（中国本周或将为空间站建设发射第二个模块。）

9. 自从着陆巡视器周六抵达火星表面以来，中国国家航天局很少公布火星探测任务的进展情况。航天局周三表示，着陆器和火星车的各部件，包括太阳能电池板，已"正常部署到位"。

10. 那张黑白照片显示，引导火星车从着陆器下滑到火星表面的坡道在火星表面投射出清晰的阴影。（远处地平线成弧线状是受广角镜头的影响。）

11. 另一张彩色照片显示了火星车的尾部以及为火星车提供动力的太阳能电池

板已成功展开的种种迹象。照片的背景是乌托邦平原的红色岩石和土壤。1976 年，美国国家航空航天局的"海盗 2 号"探测器也曾在这个撞击盆地着陆。

12. 布朗大学地质科学教授詹姆斯·海德说："这里的景象与'海盗 2 号'着陆器着陆的乌托邦盆地北侧的景象截然不同，那里有很多卵石和岩石。"

13. 海德博士说："'祝融号'周围的地形看起来更平坦，没有卵石，近处有一些有趣的岩石，但大部分是细粒的沉积物。"

14. 海德博士说："这可能是古火星海洋的沉积物，也可能是与冰川和冰相关的物质。"

15. 中国国家航天局还发布了两段短视频，视频记录了将飞行器运载到火星的环绕器与着陆器分离。

16. "祝融号"携带了一系列仪器来研究火星的地形、地质和大气。其中一个目标是了解该地区冰的分布情况，从理论上说，这可能会在将来某天帮助到访的人类维持生存。"祝融号"预计将在几天后驶离着陆器。

17. 美国国家航空航天局周三发表声明，局长比尔·纳尔逊就首批图像向国家航天局表示祝贺。

18. 纳尔逊先生在声明中说："随着火星探测器国际科学团体的壮大，美国以及整个世界期待着'祝融号'的新发现，以增进人类对红色星球的认识。"

【注释】

① 翻译中适当省译会使译文更简洁，例如此处将"Mars Landing"省译为"着陆"，若直译为"火星车着陆火星"则表意重复。

② 英语不定冠词有时可省译，此句中"a spacecraft"指上文所提探测器，若译为"一个探测器"会使句意不明。翻译中可适当调整主被动语态以使译文流畅，例如此句将原句宾语"spacecraft"选作译文主语，避免了"使探测器着陆火星"这种生硬的汉语表达。

③ 英语是演绎思维，汉语是归纳思维，因此翻译中需适当调整语序。此句若保留英语原语序，用破折号解释照片和视频的具体内容，句式会显欧化，故先译照片视频的具体内容，再以"这些照片视频"作为归纳，这样更符合汉语的思维习惯。

④ 英译汉时可适当将名词结构动词化，此句若根据英语句子结构将宾语从句译为"……的猜疑"则定语部分过长，故将"speculation"译为动词"猜疑"并增添主语"人们"。

⑤ "orbiter"：环绕器。"天问一号"环绕器具备飞行器、通信器和探测器三大功能。环绕器首先作为飞行器，将着陆巡视器送至火星着陆轨道。待成功释放着陆巡视器后，环绕器作为通信器，为着陆巡视器建立与地球之间的中继通信链路。通信工作结束后，环绕器作为探测器对火星进行科学探测。环绕器搭载的中分辨率相机、高分辨率相机、次表层探测雷达、矿物光谱分析仪、磁强计、离子与中性粒子分析仪、能量粒子分析仪等 7 台科学载荷，将获取火星形貌与地质构造、表面物质成分与土壤类

型分布、大气电离层、火星空间环境等科学数据,重点关注陨石坑、火山、峡谷、干涸河床等典型地貌和地质单元,实施高分辨率探测。

⑥ "lander":着陆器,是降落在天体表面的一种航天器,对于有大气层的天体,着陆器需要经历穿越大气层的过程。因此,这类着陆器可能需要配备降落伞来减速,将末速度控制在合理范围内,最终在天体表面软着陆。"天问一号"着陆器与环绕器分离后,需要在 9 分钟的时间里将速度从约每秒 4.9 千米降到 0。它首先借助火星大气进行气动减速,克服超高速摩擦产生的高温、气动带来的姿态偏差等挑战,将每小时约 2 万千米的下降速度减掉 90% 左右;接着打开降落伞,进行伞系减速;当速度降至每秒 100 米时再通过反推发动机减速,进入动力减速阶段;距离火星表面 100 米时,进入悬停阶段,精准避障、缓速下降后,着陆器在缓冲机构和气囊的保护下稳稳降落在火星乌托邦平原南部预选着陆区。

⑦ 根据实际情况将"July"和"February"增译为"去年 7 月"和"今年 2 月",直译会使时间不合逻辑。

⑧ 定语从句较长时可分译为以先行词作主语的小句,避免所修饰的名词前定语过长。

⑨ 此处根据背景知识将副词短语"very briefly"增译为单独的小句,苏联的"火星 3 号"1971 年 12 月 2 日成功着陆火星,工作了仅 20 秒就与地球失去了通信联系。

⑩ 英语使用代词频率高于汉语,英译汉时可适当将代词增译为所指代的名词,使句意更清晰,此句中"it"承接上文指中国国家航天局。

⑪ "Utopia Planitia":乌托邦平原,位于一个直径大约 3300 千米的撞击盆地内,是火星上最大的平原。科学家利用美国国家航空航天局"火星勘测轨道飞行器"(MRO)搭载的"浅地层雷达"(SHARAD)对该平原部分区域进行了研究,认为其表面下可能有大量的水冰储备,乌托邦平原很可能曾是一片海洋。

⑫ 英语为形合语言,介词使用频率高于汉语,英语介词不一定要译为汉语介词,例如此句将介词短语"on the northern side of the Utopia Basin"转换为形容词短语"乌托邦盆地北侧的"。

⑬ "boulderless"由 boulder 和后缀-less 构成,-less 意为缺乏,而非更少,注意应译为"没有卵石"而非"卵石更少"。

⑭ 此处介词"of"引导的"videos"后置定语中又包含了定语从句,结构较复杂,难以译为"videos"的前置定语,故采取分译,添加主语"视频",并配合主语将介词"of"译为动词"记录"。

翻译练习 4.6

Deflecting an Asteroid Before It Hits Earth May Take Multiple Bumps

1. After years of shooting meteorites with a special gun owned by NASA,

researchers highlighted challenges for a preferred method of planetary defense.

2. There's probably a large space rock out there, somewhere, that has Earth in its cross hairs.① Scientists have in fact spotted one candidate—Bennu,② which has a small chance of banging into our planet in the year 2182. But whether it's Bennu or another asteroid, the question will be how to avoid a very unwelcome cosmic rendezvous.③

3. For almost 20 years, a team of researchers has been preparing for such a scenario.④ Using a specially designed gun, they've repeatedly fired projectiles at meteorites and measured how the space rocks recoiled and, in some cases, shattered.⑤ These observations shed light on how an asteroid might respond to a high-velocity impact intended to deflect it away from Earth.

4. At the 84th annual meeting of the Meteoritical Society⑥ held in Chicago this month, researchers presented findings from all of that high-powered marksmanship. Their results suggest that whether we're able to knock an asteroid away from our planet could depend on what kind of space rock we're faced with, and how many times we hit it.

5. In the 1960s, scientists began seriously considering what to do with an asteroid on a collision course with our planet. The leading idea back then was to launch a projectile that would shatter the space rock into pieces small enough to burn up in Earth's atmosphere, said George Flynn, a physicist at State University of New York, Plattsburgh. But scientists have since come to realize that achieving such a direct, catastrophic hit is a serious challenge.

6. "It turns out, that's very hard," Dr. Flynn said.

7. The thinking is different today, and it's not the Hollywood version with a nuclear bomb, either. Rather, the current leading idea is nudging an incoming asteroid aside. The way to do that, scientists generally agree, is deliberately⑦ setting up a collision between an asteroid and a much smaller, less massive object. Known as kinetic impact deflection⑧, such a collision alters the trajectory of the asteroid ever so slightly, with the intent that its orbit changes enough to pass harmlessly by Earth.

8. "It may barely miss⑨, but barely missing is enough," Dr. Flynn said.

9. Kinetic impact deflection is a promising—and currently feasible—technique, said Dan Durda, a planetary scientist at the Southwest Research Institute in Boulder, Colo. "It doesn't require science fiction kinds of technologies."⑩

10. In 2003, Dr. Flynn, Dr. Durda and colleagues began firing projectiles at meteorites to test the limits of kinetic impact deflection. The goal was to figure out

how much momentum could be transferred to a meteorite without shattering it into shrapnel that could continue on a similar orbital path through the solar system.

11. "If you break it into pieces, some of those pieces may still be on a collision course with Earth," Dr. Flynn said.

12. Similar laboratory studies in the past have mostly shot projectiles at terrestrial rocks. But meteorites are a much better sample, he said, because they're fragments of asteroids. The hitch is getting access to them.

13. "It's hard to talk museum curators into giving you a big piece of a meteorite so you can turn it into dust,"① Dr. Flynn said.

14. Over the course of many years, the researchers amassed 32 meteorites, most purchased from private dealers. (The largest, roughly the size of a fist and weighing one pound, cost the team about ＄900.)

15. Roughly half of the meteorites belonged to a type known as carbonaceous chondrites,② which tend to be relatively rich in carbon and water. The remainder were ordinary chondrites, which typically contain less carbon. Importantly, both types are representative of the near-Earth asteroids that pose the largest risk to our planet. (Bennu is a carbonaceous chondrite.)

16. The team turned to an Apollo-era facility to test how the meteorites responded to high-speed impacts. NASA's Ames Vertical Gun Range in California was built in the 1960s to help scientists better understand how moon craters form.③ It's capable of launching projectiles at over four miles per second, far faster than a rifle.

17. "It's one of the few guns on the planet that can shoot things at the speeds characteristic of impacts," Dr. Flynn said.

18. Working within the facility's firing chamber, roughly the size of a walk-in closet, the researchers suspended each space rock from a piece of nylon string. They then pumped the chamber to a vacuum—to mimic the conditions of interplanetary space—and fired tiny aluminum spheres at the meteorites. The team launched spheres ranging in diameter from one-sixteenth to one-fourth of an inch at different velocities. Several sensors, including cameras that recorded up to 71,000 frames per second, documented the impacts.

19. The goal was to determine the point at which a meteorite stops being simply nudged by an impact and instead starts to fragment.④

20. The researchers found a significant difference in the strength of the two types of meteorites they tested. The carbonaceous chondrites tended to fragment much more readily—they could withstand receiving only about one-sixth of the

momentum that the ordinary chondrites could before shattering.

21. These results have implications for deflecting a real asteroid, the team suggests. If an asteroid richer in carbon was headed our way, it might be necessary to give it a series of gentler nudges to prevent it from breaking up.

22. "You might have to use multiple impacts," Dr. Flynn said.

23. Next year, researchers will test kinetic impact deflection on a real asteroid in the solar system for the first time with NASA's Double Asteroid Redirection Test (DART)⑮ mission. The spacecraft's target asteroid, a roughly 525-foot piece of rock known as Dimorphos,⑯ is in no danger of hitting Earth, however. The mission is expected to launch in November.

24. Laboratory investigations of kinetic impact deflection shed light on how an asteroid will respond to being impacted, said Nancy Chabot, who is the DART mission's coordination lead and was not involved in the experimental work.

25. "It's definitely important to be doing these experiments," said Dr. Chabot, who is also a planetary scientist at the Johns Hopkins University Applied Physics Laboratory.

26. The DART mission is about being prepared for what's most likely a cosmic inevitability.

27. "It's one of these things we hope we never actually need to do," Dr. Chabot said. "But the Earth has been hit by objects for its entire history, and it will continue to get hit by objects in the future."⑰

【参考译文】

或许多次碰撞才能使小行星偏离轨道避免撞击地球

1. 多年来,研究人员尝试使用美国国家航空航天局的特种火炮轰击陨石,但他们仍然特别指出,这种方法作为行星防御的优选方案仍面临诸多挑战。

2. 在宇宙某处可能有一块巨型太空岩块正朝着地球袭来。科学家们的确发现了这样的岩块——贝努。虽然这颗小行星在2182年撞上地球的可能性很小,但不管是贝努还是其他小行星,如何避免一次不必要的太空预定会合才是问题所在。

3. 近20年来,一个研究小组一直致力于防患于未然。他们用一种特殊设计的火炮,反复向陨石发射炮弹,以评估陨石受轰击后的退缩情况,甚至在某些情况下,陨石被击碎的状况。观察结果清晰地展现了在受到使其偏离轨道,绕过地球的高速撞击时,陨石会作何反应。

4. 本月,在芝加哥举行的第84届国际陨石学会年会上,研究人员展示了强力轰击陨石的全部实验结果。他们的研究结果显示,能否通过撞击,使小行星避开地球,可能取决于太空岩块的类型以及撞击次数。

5. 在20世纪60年代，科学家们开始认真考虑怎么应对一颗即将撞击地球的小行星。美国纽约州立大学普拉茨堡分校的物理学家乔治·弗林说，那时的主流观念是发射一颗炮弹，将小行星炸成小碎片，小到可以落入地球大气层燃尽。但后来科学家们意识到，这样干净利落地击毁小行星绝非易事。

6. "事实证明，这种方法非常困难，"弗林博士说。

7. 如今，观念发生转变，也不会像好莱坞电影那样，用核弹爆炸击碎小行星。相反，现在的主流观点是把飞来的小行星推到一边。科学家们普遍认为，要做到这一点，需要精心策划，让小行星与一个体积和质量小很多的物体发生碰撞。这种被称为动能撞击偏转的方式，是通过轻微地改变小行星的运行轨迹，从而使其经过地球时，不会对我们造成任何损害。

8. 弗林博士说："小行星可能刚好与地球擦肩而过，但这就足够了。"

9. 行星科学家丹·杜尔达就职于位于科罗拉多州博尔德市的美国西南研究院。他表示："动能撞击偏转技术前景广阔，而且目前可以实现。阻止小行星撞击地球并不需要科幻小说中的那些技术。"

10. 在2003年，弗林博士、杜尔达博士和同事们开始向陨石发射炮弹，以测试动能撞击偏转的极限。其目标是计算出在不粉碎陨石的前提下，到底需要多少动量才能使其沿着与原来类似的轨道继续穿过太阳系。

11. 弗林博士说："如果只是击碎小行星，部分碎片仍然有可能与地球相撞。"

12. "过去实验室中类似的研究大多是用炮弹轰击陆地岩石。作为小行星碎片的陨石无疑是更合适的实验样本，但困难在于如何获得陨石。"

13. 弗林博士说："说服博物馆馆长让出一大块陨石，让你把它打得粉碎，这事可不好办。"

14. 经过多年努力，研究人员收集到了32块陨石，其中大多数购自私人经销商。（最大的一块约有一拳头大，重达一磅，研究小组为此花了大约900美元。）

15. 约半数陨石是碳质球粒陨石。这种陨石往往含有更多的碳和水。其余的都是普通球粒陨石，一般含碳量较低。重要的是，对地球构成最大威胁的近地小行星都属于这两种类型。（贝努属于碳质球粒类陨石。）

16. 该团队使用一处阿波罗时代的设施以测试陨石对高速撞击的反应。这处设施就是美国国家航空航天局艾姆斯垂直射击靶场，建于20世纪60年代，位于加利福尼亚州。其建造目的是帮助科学家更深入地研究月球陨石坑的形成过程。这个靶场能够以超过每秒4英里（6.4千米）的速度发射子弹，比步枪快得多。

17. 弗林博士说："这把枪能以高速撞击速度射击物体。地球上这样的枪支很少见。"

18. 在大约与步入式壁橱一样大小的发射室内，研究人员先将陨石悬挂在尼龙绳上，然后将室内抽成真空——以模拟太空环境——随后向陨石发射小铝球。研究小组以不同的速度发射了小球，小球直径范围 1/16 英寸（0.001 59 米）到 1/4 英寸（0.006 35 米）。现场的几个传感器（包括帧率高达每秒 7.1 万帧的摄像机）记录下了撞击画面。

19. 实验目标是确定陨石停止被撞击推动那一刻到陨石破碎成片之间的临界点。

20. 研究人员发现，测试的这两类陨石的强度存在显著差异。碳质球粒陨石更容易碎裂——其承受的动量只有击碎普通球粒陨石动量前的 1/6。

21. 研究小组认为，这些研究结果在使真正的小行星偏离轨道时有借鉴意义。如果来袭的小行星含碳量丰富，可能需要一系列轻一些的撞击，以防其解体。

22. 弗林博士表示："为此，我们可能需要多次撞击。"

23. 明年，在美国国家航空航天局的双小行星重定向测试任务中，研究人员将会首次对太阳系中的一颗真正的小行星进行动能撞击偏移测试。但是，此次任务中航天器的目标小行星迪莫福斯是一块直径约 525 英尺（160 米）的陨石，它不会撞击地球。该任务预计将于今年 11 月启动。

24. 南希·夏博特是双小行星重定向测试任务的协调负责人，她没有参与实验工作。她表示："在实验室中的动能撞击偏移研究揭示了小行星在受到撞击时将做出什么反应。"

25. 夏博特博士也是美国约翰·霍普金斯大学应用物理实验室的行星科学家。她说："这些实验非常重要。"

26. 双小行星重定向测试任务就是要对宇宙中最有可能出现的不可避免的情形做好准备。

27. "我们真心希望永远不用启动类似这样的任务，"夏博特说，"但从历史来看，地球之前受到过外来天体的撞击，未来还会继续受到其他天体的撞击。"

【注释】

① 从句强调"地球在太空岩块的轨迹上"，但主句的主体是"太空岩块"，为了还原原文中想要强调的"太空岩块飞向地球"这一重点，翻译时要调整表达方式，使前后一致。

② "Bennu"：贝努，一颗近地小行星。

③ "a very unwelcome cosmic rendezvous"指的是小行星将会与地球相撞这件事，在理解其含义基础上可以保留原文形式。而"rendezvous"指的是事先约好的会面，所以翻译时要在"会合"前加上"预定"。

④ "prepare for such a scenario"指的是为防止小行星与地球碰撞做准备，汉语中可以使用有"防范灾祸"含义的成语或俗语来表达。

⑤ 此处的"how"可以翻译为汉语中的范畴词"……的情况"。

⑥ "Meteoritical Society"：国际陨石学会。

⑦ 此处指的是人为制造碰撞，不可直译成"故意地"，可以将其译为"精心策划"。

⑧ "kinetic impact deflection"：动能撞击偏转。

⑨ 可以适当运用汉语四字格，使文章更加连贯，提高其可读性。

⑩ 本句中的"it"指的是阻止小行星撞击地球。

⑪ 此句中的"it's hard to…"可以单独提出来，放在句末，这样汉语句子不会过长，便于读者阅读。

⑫ "carbonaceous chondrite"：碳质球粒陨石；"ordinary chondrite"：普通球粒陨石。

⑬ 此句中含有信息集中，包含隶属机构、名称、建成时间、地点、目的等，翻译时可以将机构和名称合并，按照名称、时间、地点、目的排序，并适当断句。

⑭ 实验目的是要推动陨石，改变其行动轨迹，而不是把它击碎，所以"the point"指的是临界点。

⑮ "Double Asteroid Redirection Test（DART）"：双小行星重定向测试。

⑯ "Dimorphos"直径约160米（约520英尺），所以525英尺指的是直径。

⑰ 根据文章大意，此句中的"object"指的是小行星一类的天体。

翻译练习 4.7

Who Gets Sick in Space? Orbital Tourists May Offer Better Clues

1. Commercial spaceflight brings a more diverse cross section of humanity beyond Earth, helping medical researchers collect data.

2. Doctors have poked and prodded NASA astronauts for years, and the astronauts, as government employees, have largely acceded to their roles as test animals in studying how an alien environment—outer space—affects the human body①.

3. But professional astronauts have historically been a small slice of humanity. Initially, they were chosen from the ranks of military test pilots who were white, physically fit men. Later, as NASA broadened its criteria, it still chose only astronauts who passed their physical thresholds.

4. But that could be changing as private spaceflight opens space to a more diverse cross section of humanity②.

5. For scientists, the change will create a bonanza of new data on how the human body adapts to space.

6. The Inspiration4 mission, which launched on Wednesday, shows how medical researchers can benefit from the new commercial space travel③. The crew of four, none of whom is a professional astronaut, will spend a fair amount of time in orbit helping to advance medical research.

7. One of the passengers, Hayley Arceneaux, exemplifies those possibilities. At 29, she is younger than most space travelers, a cancer survivor and will be the first person in space with a prosthesis—metal rods that were implanted after a tumor was removed from her left leg④.

8. "We're going to learn some things that are very fundamental," said Dorit Donoviel, executive director of the Translational Research Institute for Space Health, or TRISH, at the Baylor College of Medicine in Houston⑤, which is coordinating research during the Inspiration 4 flight.

9. Research to date has revealed that fluids shift upward in the body in the absence of gravity—swollen heads, shrunken legs. The lack of gravity also weakens bones. Not only does radiation inspace slam into DNA creating mutations, but the unusual conditions of weightlessness cause some genes to turn on and others to turn off⑥. The biological repercussions of these alterations are not yet understood.

10. The crew aboard Inspiration4 will take 10 tests originally designed to gauge the mental performance of NASA astronauts each day. The tests take about 20 minutes to complete.

11. "It needed to be brief, because astronauts kind of hate doing these things," said Dr. Mathias Basner, a professor of psychiatry at the University of Pennsylvania who is the lead investigator for that experiment.

12. But in a dangerous environment like space, small errors can lead to catastrophe.

13. "So we need our astronauts to perform at their best all the time," Dr. Basner said. "Now, the problem is that humans are particularly bad at self-assessing their performance capability, especially in chronic exposure situations⑦. If you sit in the same environment all the time, you think you'll be fine, but you're actually not."

14. One test is simply a square that appears on a screen, and one has to tap on it. The square changes position and progressively becomes smaller. That measures the reaction speed and eye-hand coordination.

15. Another measures what is known as psychomotor vigilance. First, the subject stares at a box on the screen. A stopwatch suddenly pops up within the box, counting the milliseconds until the subject presses a button. "Which is extremely sensitive to sleep deprivation," Dr. Basner said.

16. Yet another test gauges a person's ability to identify emotions in other people.

17. The test displays 20 faces showing a variety of emotions—happy, sad, angry, fearful or expressing no emotion. In a so-called bed rest study—lying down for extended periods of time mimics many of the physical effects of weightlessness in space—subjects were able to still correctly identify most of the emotions. But they took longer to identify them, and their responses skewed to more negative expressions.

18. Mark J. Shelhamer, a professor at Johns Hopkins Medicine, is gathering data on how spaceflight affects the vestibular system—the parts of the human body, particularly the inner ear, that maintain balance.

19. His research consists of two parts, to be conducted before launch and after the return to Earth. One will measure the crew members' postures. "It's just what it sounds like," Dr. Shelhamer said. "It's the ability to stand up. And the ability to stand up is not only based on muscle strength, but it's based on coordination."

20. The Inspiration4 crew members will hold a Windows tablet to their chests, put their feet together and close their eyes. Accelerators in the tablet will measure how much they sway as they stand.

21. "That's not so hard to do on Earth," Dr. Shelhamer said. "But it can be challenging after having spent some time in space."

22. Dr. Shelhamer has also devised a test using the tablet to study whether weightlessness causes eyes to become misaligned. That could give hints of how the brain might become confused, disrupting the sense of balance.

23. What researchers want to learn is how to predict who becomes sick in space. Surprisingly, there is no correlation between who experiences motion sickness on Earth—on a boat rocking back and forth, during a long car trip, even short episodes of floating during parabolic plane flights—and those who become sick in orbit[⑧].

24. "We don't understand this at all," Dr. Shelhamer said. "It's an embarrassment[⑨], because we would like to think that it's the same fundamental mechanisms—the fact that the different sensory systems don't match up."

25. TRISH, the organization Dr. Donoviel leads, has built a database to store the study results that not only protects the privacy of private space travelers in accordance with the requirements of HIPAA, the Health Insurance Portability and Accountability Act, but will also allow researchers to compare data from different studies for one individual.

26. For instance, Dr. Shelhamer might want to know whether an individual whose ability to stand straight without swaying declined after a trip to space also

encountered cognitive slowdowns while taking Dr. Basner's tests in space. That is often not possible with how NASA reports data collected on its astronauts，he said⑩.

27. In addition，the Inspiration4 crew members are wearing Apple Watches that will measure their heart beats and oxygen levels. They will also test out ultrasound devices that can track how water in their bodies shifts upward while they float in orbit. That could help solve the puzzle of the squashing of eyeballs and resulting shift in vision experienced by some astronauts.

28. The research does⑪ depend on how well the Inspiration4 crew feels during their trip. Because there is no way to predict who gets sick in space，it is possible that they all will and then they may not be able to do much research during the three-day trip.

29. "My worry is that they're not going to enjoy their time in space，" Dr. Donoviel said. "They may not be able to carry out the experiments that we are sending up. So that is a risk that I'm taking. "

【参考译文】

谁会在太空中生病？太空环游旅行者可能会提供更好的线索

1. 商业太空飞行将更多不同类型的人带入太空，有助于医学研究人员收集数据。

2. 多年来，医学人员一直在探查美国国家航空航天局航天员的身体变化，而作为政府雇员的航天员们，也在很大程度上接受了他们作为实验对象的角色，协助研究外星环境即外太空如何影响人体。

3. 但职业航天员历来只是人类的一小部分群体。最初，他们是从身体健康的白人男性军事试飞员中选出的。后来，尽管美国国家航空航天局放宽了标准，但仍然只选择身体条件达标的航天员。

4. 但随着私人太空飞行向更多不同类型的人群开放，情况可能发生变化。

5. 对科学家们来说，这一变化将为研究人体如何适应太空带来海量新数据。

6. 周三发射的"灵感4号"飞船表明，这种新型商业太空旅行有助于医学人员的研究工作。飞船上的四名机组人员皆为非职业航天员，他们将投入大量时间沿轨道环行，协助推进医学研究。

7. "灵感4号"上有一名乘客叫海莉·阿塞尼奥，她本人就印证了上述种种可能性。她今年29岁，比大多数太空旅行者都要年轻；她曾得过癌症，在切除左腿上的肿瘤后安装了金属义肢，因此她将成为第一个戴义肢进入太空的人。

8. 美国休斯敦贝勒医学院空间健康转化研究所的执行主任多利特·多诺威尔

说:"我们会了解到一些非常基础的东西"。该研究所在"灵感4号"飞行期间负责协调研究工作。

9. 最新研究表明,在没有重力的情况下,体液会上涌,导致头部膨胀,腿部萎缩。重力缺失也会使骨骼变得脆弱。此外,太空辐射不仅会冲击 DNA 引起突变,失重的异常环境还会导致一些基因活跃、一些基因呆滞。这些变化对生物机体有何种影响目前尚不清楚。

10. "灵感4号"上的机组人员每天用大约20分钟来完成10项测试。这些测试原本用来评估美国国家航空航天局航天员的精神状态。

11. "测试必须简短,因为航天员们不太喜欢做这些事情,"宾夕法尼亚大学精神病学教授马西尔斯·贝斯纳博士说。他是这项实验的首席研究员。

12. 但在太空这样的危险环境中,小差错也可能酿成大祸。

13. "所以我们需要航天员一直保持最佳状态,"巴斯纳博士说。"但问题是人们尤其不善于自我评估表现情况,特别是长期身处某些环境时。如果一直处于同一种环境,则有可能对自己的身体状况产生误判。"

14. 其中一项测试是,屏幕上就只出现一个正方形,需要受试者点击。正方形会不断变换位置并逐渐变小。这项测试旨在检验航天员的反应速度和手眼协调能力。

15. 另一项测试检验所谓的精神运动警觉性。首先,受试者盯着屏幕上显示的一个方框。方框里突然出现秒表计时,测算出受试者按下按钮作出反应的毫秒数。巴斯纳博士说,"这项测试能极其灵敏地判定航天员的睡眠缺失情况"。

16. 还有一项测试评定航天员识别他人情绪的能力。

17. 测试会展示20张表情各异的面孔,或高兴、或悲伤、或愤怒、或害怕,或面无表情。研究人员通过超长时间躺卧模拟太空失重对身体造成的诸多影响,这就是所谓的卧床休息研究。研究发现,尽管受试者仍能正确地识别大多数情绪,但耗时更长,而且他们更易识别出消极情绪。

18. 约翰·霍普金斯医学院的马克·舍尔哈默教授正在收集有关太空飞行如何影响前庭系统的数据。前庭系统是人体内耳的一部分,负责保持人体平衡。

19. 他的研究包括两部分,分别在发射前和返回地球后进行。第一部分评估机组人员的站立姿态。"就是字面意思,"舍尔哈默博士说,"测试他们的起身站立能力。起身站立不仅需要肌肉力量,还需要协调能力。"

20. "灵感4号"机组人员将一个平板电脑抱在胸前,双脚并拢,闭上双眼。平板电脑中的加速装置可测量他们站立时的摇摆程度。

21. "这在地球上并不难做到,"舍尔哈默博士说,"但在太空待了一段时间后,再做这些动作就有点困难了。"

22. 舍尔哈默博士还设计了一项测试,用平板电脑来研究失重是否会导致眼睛错位。这项测试可能为研究大脑如何变得混乱,干扰平衡感提供一些启示。

23. 研究人员想要了解的是如何预测谁会在太空中生病。在地球上,有人会因为船体前后摇晃、长时间乘车以及飞机短暂颠簸起伏而出现晕车、晕船和晕机情况，还有人因太空轨道环行而出现类似症状。令人惊讶的是，在地球和太空中出现的此类相似症状并无关联。

24. "我们完全不明白为什么会这样，"舍尔哈默博士说，"这很费解，因为我们倾向于认为这些症状的基本机制是相同的，但事实是不同的感官系统并不匹配。"

25. 在多利特博士的领导下，空间健康转化研究所创建了一个数据库来存储这些研究成果。数据库不仅按照健康保险隐私及责任法案的要求保护私人太空旅行者的隐私，还为研究人员比对源于同一个体的不同研究数据提供了便利。

26. 例如，舍尔哈默博士可能就想知道，一个人在太空旅行后出现站立不稳的现象，他是否在之前的巴斯纳博士太空测试过程中也表现出认知迟钝。多利特博士说，美国国家航空航天局收集其航天员数据的方式通常满足不了舍尔哈默博士的需求。

27. 此外，"灵感4号"机组人员还佩戴了苹果手表以测量他们的心跳和血氧。他们还将用超声波设备来追踪太空轨道环行漂浮时体液向上涌的情况。这些措施可能有助于解决某些航天员眼球受压以及由此带来的视觉变化这一难题。

28. 毫无疑问，这项研究取决于"灵感4号"机组人员在太空旅途中的身体状况。因为无法预测谁会在太空中生病，很可能所有人都生病，那么在这种情况下，三天的旅程可能就做不了太多研究了。

29. "我担心他们在太空中的状态不会太好，"多诺威尔博士说，"或许无法开展我们发送上太空的那些实验。这就是目前我承担的风险。"

【注释】

① 遇到破折号中解释前文的内容时，可译为"即"，并在汉语表述中省略破折号。

② "cross section"这里的意思是（具有代表性的）人群，"a more diverse cross section of humanity"译为"更多不同类型的人群"，表达贴切、准确。

③ 该句为表达难点，为突出句子重点，翻译时应将从句的主语与宾语对调"benefit from"译为"……有助于……"。

④ 此句将被动句译为主动句，调整"海莉"为句子主语，"metal rods that were implanted after a tumor was removed from her left leg"译为"在切除左腿上的肿瘤后安装了金属义肢"，句子更加通顺流畅，符合汉语习惯。

⑤ 翻译时应注意补充地点、时间等背景信息"Baylor College of Medicine in Houston"，此处补充地点背景"美国"，译为"美国休斯敦贝勒医学院"。

⑥ "turn on"与"turn off"若直译为"打开"与"关闭"，则过于生硬，在描述"gene"时应采用意译方法，译为"活跃"和"呆滞"，使表达更加顺畅。

⑦ "in chronic exposure situations"为理解难点，"chronic"意为"长期的、短期难以消除的"，带有负面含义，例如：a chronic worry（长期忧虑），chronic unemployment（长期失业）。此处可译为"长期身处某些（不利）环境时"或"长期暴露于某些（不利）环境条件下"。

⑧ 此句为表达难点,先翻译"there be"句型后面的从句作主语,再翻译"there be"句型的主句。注意翻译"correlation"时要明确指出是"哪两者"无关联,避免指代不明。

⑨ "embarrassment"原意为"尴尬""难堪",联系上下文,此处译为"费解"。

⑩ "That is often not possible"若直译为"通常不可能实现"则过于生硬,因此应结合上下文内容,译为"满足不了需求",使表达更加通顺。

⑪ 此处"does"可译为"毫无疑问",表示强调,翻译时应注意不要漏译。

翻译练习 4.8

NASA's Perseverance Rover① Stashes First Mars Rock Sample

1. The rock, sealed in a tube, is the first of many the robotic explorer will collect to one day send back to Earth for scientists to study.②

2. NASA's Perseverance rover on Mars has confirmed the successful collection of its first rock sample.

3. "One down, a lot more to go!" Kenneth A. Farley, professor of geochemistry at the California Institute of Technology and the mission's project scientist, said in an email.

4. On Monday night, NASA announced that the rover had sealed the tube that contains the rock core, which is slightly thicker than a pencil, and put it away for safekeeping in its belly. That and other collected samples will③ be dropped to the ground to be collected by another spacecraft. They will eventually be ferried back to Earth,④ helping scientists study the mysteries of the red planet in much the same way that lunar samples from the Apollo and Soviet missions helped advance understanding of the moon.⑤

5. For decades, planetary geologists have wanted to study rocks from another planet. They have done that with pieces of Mars that were blasted into space by meteor impacts and later passed Earth's path and landed as meteorites.⑥ But with Martian meteorites, scientists had no idea where the rocks originated—intriguing pieces from a jigsaw puzzle but no opportunity to find the other pieces.⑦

6. With the Perseverance samples, scientists will know exactly which rocks the samples came from,⑧ and the rover will have performed detailed study of the surrounding geology.

7. The rover drilled the core from a flat, briefcase-size Mars rock nicknamed Rochette⑨ last week. In the first pictures that the rover took of the collection tube, the rock sample could be easily seen.

8. But viewers on Earth were concerned when the rock couldn't be seen⁑ in subsequent photographs. Those were taken after Perseverance used its robotic arm to shake the tube to help the sample settle to the bottom of the container. Mission managers were confident it⑪ was just hidden in shadow, but after a failed drilling attempt last month, they wanted to make sure before sealing the tube.

9. Subsequent photographs with better lighting showed the rock was still there.

10. Rochette looks to be a piece of a hardened lava, which can be precisely dated. Thus, scientists will be able to determine how old this boulder is, and it helps pin down the ages of lower, older layers.

11. In the first drilling attempt in August, everything on Perseverance appeared to function flawlessly yet the tube ended up empty. Analyzing what happened, scientists and engineers working on the mission concluded⑫ that the first rock was just too fragile and the sample fell out.

12. One of the key tasks for Perseverance, which arrived on Mars in February, is to collect rocks and soil that will eventually be brought back to Earth by another mission so that scientists can exhaustively study them using state-of-the-art instruments in their laboratories. Scientists hope to collect more than 30 samples from a variety of locations in Jezero crater, a landing site that was chosen because scientists on Earth felt an ancient river delta along the crater rim was a promising target for fossilized microbial life⑬ if it ever existed.

13. Perseverance is collecting the rock and soil samples but has no way to deliver them back to Earth. That will await the future mission, which is still being designed.

14. On Saturday, Ingenuity, the robotic helicopter accompanying Perseverance, made its 13th flight, scouting a series of outcrops to provide help in planning upcoming drives by the rover.

15. Perseverance is not the only new rover to start exploring Mars this year. China also successfully landed a rover, Zhurong, in May, and its mission, intended to last three months, has been extended. Last week, the Chinese space agency released a 360-degree panorama taken by Zhurong.

【参考译文】

美国国家航空航天局"毅力号"火星车成功采集首块火星岩石样本

1. "毅力号"采集的首块岩石样本封存于试管中,未来会连同此后采集的多个样本一起送回地球供科学家研究。

2. 美国国家航空航天局"毅力号"火星车已经确认，他们首块岩石样本采集成功。

3. "一块样本到手，未来更多不在话下！"加州理工学院地球化学教授兼探测项目科学家肯尼斯·法利在一封电子邮件中说。

4. 美国国家航空航天局周一晚宣布，火星车已将这块略粗于铅笔的岩芯装入试管密封并置于舱内，妥善保管。日后，这块岩石将会和收集的其他样本一起投放到火星表面，再由另一个航天器取走，最后运回地球。阿波罗计划和苏联登月计划采集的月球样本曾一度增进了人类对月球的了解，这些样本同样有助于科学家探秘火星这颗红色星球。

5. 几十年来，研究另一星球上的岩石一直是行星地质学家的愿望。曾因流星撞击火星发生爆炸而抛入太空的火星碎片，后来飞经地球坠落成为陨石。科学家对这些陨石进行了研究，但对其来源区域却一无所知 —— 如同拿到几个神秘的拼块，却无从找到其余拼块完成拼图一样。

6. 而对于这次采集的样品，科学家会确切知道它取自哪些岩石，"毅力号"火星车也会对岩石周围的地质状况做详细勘察。

7. 上周，火星车在一块叫作"罗切特"的火星岩石上钻取了岩芯。这块岩石形状扁平，大小类似公文包。从火星车拍摄的采集管首批照片中，可以清晰地看见岩石样本。

8. 但在随后拍摄的照片中，却看不见岩石样本，这引发了地面观察者的担心。"毅力号"用机械臂摇动试管以便样本能安落在试管底部，这些照片拍摄于这之后。探测计划负责人相信，岩石只是被阴影遮住了，但经历了上个月钻探失败后，这次他们想确认后再密封试管。

9. 随着光线好转，后续照片显示岩石仍在管内。

10. "罗切特"看起来是一块凝固的熔岩，其形成年代可以精确测定。因此，科学家可以确定这块巨石的年龄，进而测定年代更久远的深层岩层的年龄。

11. "毅力号"在八月份进行首次钻探时，所有部件似乎都运行完美，但最后试管却空无一物。负责探测计划的科学家和工程师分析后得出结论，那次采集的岩石质地太松脆，样本掉落在试管外面了。

12. "毅力号"于二月抵达火星，其主要任务之一是收集岩石和土壤样本，最后由另一航天器带回地球。这样，科学家就可以利用实验室最先进的设备对其进行全面详尽的研究。"毅力号"的着陆点选择在杰泽罗火山口。科学家认为处于这个火山口边缘的一个古老河流三角洲可能会有微生物化石(如果火星的确曾存在微生物)。科学家希望能从火山口附近的不同地点收集30多个样本。

13. "毅力号"目前正在收集岩石和土壤样本，但还无法将其运回地球，因为能完成运送任务的另一项太空飞行计划尚处于谋划阶段。

14.周六，"毅力号"搭载的火星无人机"机智号"进行了第13次飞行，侦察了一系列的岩石出露层，为规划火星车接下来的行驶路线提供参考。

15."毅力号"不是今年唯一一个开始探索火星的新探测车。五月，中国的"祝融号"火星车也成功着陆火星。"祝融号"的设计探测期限为三个月，现已超期工作。上周，中国国家航天局发布了一张"祝融号"拍摄的360度全景图。

【注释】

① "rover"在航天领域指行驶于地外天体的探测车，此处根据语境译为"火星车"。

② 此句中"many"为代词，代指岩石样本，其后的定语较长，按英语句子结构译为"……的许多岩石"会使汉语句子显得头重脚轻，故译文将此句分译为两句。翻译含多个修饰成分的句子时可调整各成分的位置和语法层级，例如此句将"sealed in a tube"这一后置定语调整为第一小句的谓语，将"is the first"这一表语调整为前置定语，将"to one day send back"这一状语调整为第二小句的谓语。

③ 译入汉语时添加时间副词"日后"会让句意更清晰，表明从"投放"到取走在较长一段时间以后。

④ 处理句子时，可采用合译法，例如此句主语为"they"，承接上文指代这一样本和其他样本，所以可以不用另起一句，避免重复主语。

⑤ 翻译时，可根据上下文适当增译使句意更清楚，具体到此句，翻译"red planet"时增译"火星"，翻译"Soviet missions"时增译"登月"，翻译"understanding of the moon"时增译"人类"。

⑥ 此句中"pieces of Mars"的定语从句较长，有"were blasted""passed"和"landed"三个谓语，可分译为单独的小句。

⑦ 英语原文破折号后为省略句，只保留名词词组，但译入汉语时需要补充谓语，避免延续原来的名词性结构，导致汉语表达生硬。

⑧ 根据上下文可知岩石样本由火星车采集，因此"came from"根据语境译为"取自"，而非宽泛的基础义项"来自"，这样汉语译文语言更地道。

⑨ 此句多个定语译为"……的火星岩石"结构会导致汉语前置定语过长，故采取分译法。

⑩ 英语的被动结构在翻译时，大多可转换为主动结构，例如译为汉语无主句或增添主语，这样可以避免保留被动结构而使汉语表达生硬，例如此句将"岩石样本不能被看见"转换为无主句"看不见岩石样本"。

⑪ 英语代词使用频率高于汉语，英译汉时可适当将代词所指具体化，使句意更清晰，此句的"it"指代岩石样本。

⑫ 翻译时可适当调整状语语序，避免受英语句型的影响，例如此句将状语"analyzing what happened"与主句谓语"concluded"结合。

⑬ 英语呈静态，倾向于多用名词，而汉语呈动态，多用动词，因此英译汉时可将

名词结构动词化，此句若保留英语名词结构，直译为"是微生物化石的可能目标"则语言生硬，故动词化为"可能会有微生物化石"。

翻译练习 4.9

Japan's Journey to an Asteroid Ends with a Hunt in Australia's Outback

1. The Hayabusa2 mission cementsJapan's role in exploring the solar system, but finding its asteroid cargo presented one last challenge.

2. This past weekend, Japan's space agency concluded a six-year, 3. 25 billion-mile journey of discovery that aims to shed light on the earliest eons of the solar system and possibly provide clues about the origins of life on Earth.

3. But first, it had to go on a scavenger hunt in the Australian outback.

4. Bits of an asteroid landed in a barren region near Woomera, South Australia. These were being ferried to Earth by Hayabusa2, a robotic space probe launched by JAXA, Japan's space agency, in 2014 to explore an asteroid named Ryugu, a dark, carbon-rich rock a bit more than half a mile wide. [①]

5. "I'm home," Yuichi Tsuda, the mission's project manager said in translated comments during a news conference after a capsule containing the asteroid sample was recovered. "Hayabusa2 is home. "

6. Dr. Tsuda described the condition of the capsule, which set down amid bushes in the Australian desert, as "very perfect. "

7. The success of the mission and the science it produces will raiseJapan's status as a central player in deep space exploration, together with [②] NASA, the European Space Agency and Russia. JAXA currently has a spacecraft in orbit around Venus studying that planet's hellish climate and is collaborating with the Europeans on a mission that is on its way to Mercury.

8. In the coming years, Japan plans to bring back rocks from Phobos, a moon of Mars, and contribute to NASA's Artemis program to send astronauts to Earth's moon.

9. But the immediate challenge was finding the 16-inch-wide return capsule somewhere amidthousands of square miles in a region 280 miles north of Adelaide, the nearest large city.

10. "It's really in the middle of nowhere," [③] Shogo Tachibana, the principal investigator in charge of the analysis of the Hayabusa2 samples, said in an interview. He is part of a team of more than 70 people from Japan who traveled to Woomera for recovery of the capsule. The area, used by the Australian military for

testing, provides a wide open space that was ideal for the return of an interplanetary probe.

11. The return capsule separated from the main spacecraft on Saturday about 12 hours before the landing, when it was about 125,000 miles from Earth.④ The mission's managers confirmed the capsule's ejection using data beamed back from the spacecraft, as well as with visual assistance from telescopes, like one at Kyoto University in Japan.⑤

12. Soichi Noguchi, a Japanese astronaut who joined the International Space Station crew in November after a trip in a SpaceX capsule, said he spotted⑥ Hayabusa2 from orbit.

13. A day before, heavy rains fell and strong winds blew in Woomera. But then the air cleared and calmed. "The weather was crystal clear," Dr. Tsuda said.

14. The capsule was spotted re-entering the atmosphere during the pre-dawn hours of Australia on Sunday, a tail streaming behind it as the atmosphere heated its surface.

15. Minutes later, the mission's managers detected a radio signal from a beacon in the capsule.

16. From the fireball, the radio beacon and radar readings, the recovery team calculated the likely landing spot, and a helicopter was sent from Woomera to search for the capsule, although it was still dark.

17. "I was very, very, very nervous and uneasy," Satoru Nakazawa, a project sub-manager who was part of the Woomera recovery team, said during the news conference.

18. But then, soon after sunrise, the capsule and its parachute were spotted. "We thought, wow, we found that," Mr. Nakazawa said.

19. Even with the capsule in hand, there is a bit of a rush.⑦ The team⑧ wanted to whisk it back to Japan within 100 hours after the landing. Even though the container is sealed, the worry is that Earth air will slowly leak in. "There is no perfect sealing," Dr. Tachibana said.

20. The helicopter took the capsule to a laboratory that has been set up at the Australian air force base at Woomera. There an instrument extracted gases within the capsule that may have been released by the asteroid rocks as they were shaken and broken during re-entry.

21. Makoto Yoshikawa, the mission manager,⑨ said in an interview the scientists would also like to see if they can detect any solar wind particles of helium that slammed into the asteroid and became embedded in the rocks.

22. The gases would also reassure the scientists that Hayabusa2 did indeed successfully collect samples from Ryugu. A minimum of 0.1 grams，or less than 1/280th of an ounce，is needed to declare success. The hope is the spacecraft brought back several grams.⑩

23. On Monday night，an airplane left⑪ Australia to carry the sample back to Japan. There，the Hayabusa2 team will examine the Ryugu samples in earnest. In about a year，some of the samples will be shared with other scientists for additional study.

24. To gather these samples，Hayabusa2 arrived at the asteroid in June 2018. It executed a series of investigations，each of escalating technical complexity. It dropped probes to the surface of Ryugu，blasted a hole in the asteroid to peer at what lies beneath and twice descended to the surface to grab small pieces of the asteroid，⑫ an operation that proved much more challenging than expected because of the many boulders on the surface.

25. Studying water trapped in minerals from Ryugu could give hints if the water in Earth's oceans came from asteroids，and if carbon-based molecules could have seeded the building blocks for life.

26. Part of the Ryugu samples will go to NASA，which is bringing back some rocks and soil from another asteroid with its OSIRIS-REX mission. The OSIRIS-REX space probe has been studying a smaller carbon-rich asteroid named Bennu and it will start back to Earth next spring，dropping off its rock samples in September 2023.

【参考译文】

日本小行星探测之旅，在澳大利亚内陆地区告终

1. "隼鸟 2 号"行星探测任务巩固了日本在太阳系探索方面的地位，但他们还面临最后一个挑战，即找到从小行星返回的太空舱。

2. 上周末，日本航天局结束了一个为期六年、飞行距离长达 32.5 亿英里（52 亿千米）的研究项目。此项探索之旅的目的是揭示出太阳系最早存在的时代，并为探索地球的生命起源提供线索。

3. 但首先，日本航天局必须在澳大利亚内陆地区进行地毯式搜索。

4. "隼鸟 2 号"是日本宇航局于 2014 年发射的一个机器太空探测器，用于探索一颗名为龙宫的小行星。那是一颗漆黑的、富含碳元素的岩质行星，其直径略大于半英里。之前这颗小行星的碎片由"隼鸟 2 号"运送，落在南澳大利亚伍默拉附近的一片荒地上。

5. 在找回装有小行星样本的太空舱后,日本举行了一场新闻发布会。项目负责人津田雄一在发布会上发表评论,翻译过来就是:"我回家了,'隼鸟2号'回家了。"

6. 返回舱降落在澳大利亚沙漠的灌木丛中,津田博士称它的状况"相当完美"。

7. 此次成功及其产生的科学成果使日本在外太空探索方面的地位得到巩固,可以比肩美国国家航空航天局、欧洲航天局和俄罗斯。目前,日本宇航局的一个在轨探测器正在绕金星飞行,目标是研究该星球地狱般炎热的气候。同时他们正在与欧洲人合作,执行一项前往水星的太空飞行任务。

8. 未来几年,日本还计划从火星的卫星"火卫一"上带回岩石样本,同时帮助美国国家航空航天局完成阿尔忒弥斯计划,将航天员送往地球的卫星。

9. 但眼前还有一个挑战,那就是在数千平方英里的区域中找到16英寸(0.41米)宽的返回舱。这片荒野以北280英里(451千米)处的阿德莱德是距离它最近的大城市。

10. "它真的是荒郊野外。"负责"隼鸟2号"样品分析的首席调查员橘正吾在接受采访时说。日本派出70多人前往伍默拉回收太空舱,他就是其中一员。曾是澳大利亚军方的测试场的这片区域占地开阔,是星际探测器返回时理想的着陆点。

11. 周六,在距离地球约12.5万英里(20.21万千米)时,返回舱与主航天器分离。12小时后,它在地球着陆。该飞行任务的负责人根据从航天器传回的数据以及望远镜(如日本京都大学望远镜)观测,确认太空舱已弹出。

12. 日本航天员野口总一搭乘美国太空探索技术公司研发的太空舱,于11月开始在国际空间站工作。他称自己曾从轨道上看到了"隼鸟2号"。

13. 太空舱返回的前一天,伍默拉大雨倾盆,狂风阵阵。但后来雨过天晴,风也停了。津田博士说:"这边的天气非常晴朗。"

14. 周日,澳大利亚天亮之前的那几个小时,有人发现该太空舱重新进入了大气层。太空舱表面与大气层摩擦后,温度升高,在其身后的空中留下长长的尾巴。

15. 几分钟后,该任务的负责人检测到了来自太空舱内某个信标台的无线电信号。

16. 根据火球、无线电信标和雷达的读数,太空舱回收小组计算出了其可能的着陆点。尽管当时天还没亮,他们还是从伍默拉派遣了一架直升机进行搜寻。

17. 伍默拉回收组成员、项目分管负责人中泽悟在新闻发布会上说:"我当时非常紧张和不安。"

18. 但日出后不久,就发现了太空舱及其降落伞。中泽先生说:"我们想,哇!可算找到它了。"

19. 即使已经成功找到了太空舱,但时间还是比较紧张。回收小组希望在太空舱着陆后100小时内将其迅速送回日本。尽管太空舱已经密封,他们还是担心地球上的空气会慢慢渗入。"我们无法将它彻底封死,"橘博士说。

20. 直升机将太空舱带到了位于伍默拉的澳大利亚空军基地的实验室中。在那里可以用一台仪器抽取太空舱内的气体。这些可能是小行星岩石在重返大气层时被震碎后溢出的气体。

21. 回收小组负责人吉川诚在接受采访时说,科学家们还想看看他们是否能够探测到撞向小行星又嵌入岩石中的太阳风氦粒子。

22. 这些气体使科学家们彻底放心了,"隼鸟2号"确实成功地从龙宫(Ryugu)采集了样本。不过至少要带回0.1克,或不到1/280盎司的样本才算真正胜利。他们希望太空舱已经带回了几克。

23. 周一晚,一架飞机从澳大利亚起飞,带着样本返回了日本。"隼鸟2号"团队将在日本国内仔细研究龙宫样本。约一年后,他们将与其他科学家分享其中一些样本,以便进行其他方面的研究。

24. 为了收集这些样本,"隼鸟2号"于2018年6月就飞抵了该小行星。它先是进行了一系列调查,其技术难度不断增大。随后它向龙宫表面投放了探测器,又在其表面炸出一个坑来观察这个小行星地下的情况。此外,为了获取小行星上的碎块,"隼鸟2号"还两次降落到其表面。由于这个小行星布满巨石,降落其上比预期的更困难。

25. 如果想搞清楚地球上海洋中的水是否来自小行星,碳基分子又是否是构成生命的基本要素,那么从龙宫矿物质中所含的水分研究总可以得到些启示。

26. 部分龙宫样本将送往美国国家航空航天局。美国国家航空航天局正通过其源光谱释义资源安全风化层辨认探测器研究项目从另一颗小行星带回的一些岩石和土壤。该探测器一直在研究一个更小的富含碳元素的小行星,名为本努。它将在明年春天启程返回地球,于2023年9月带回本努的岩石样本。

【注释】

① 本句在翻译时没有按顺序先译小行星碎片,而是先处理其后的同位语部分,解释"隼鸟2号"探测器及小行星。对背景及术语进行解释后再提小行星碎片就更加顺理成章了。

② "together with"本义为"与……一起",即与下述三个国家在此领域的地位等同,因此可以意译为"与……比肩"更加妥当通顺。

③ "middle of nowhere":荒郊野外。

④ 本句在翻译时需调整语序并分译。先译出时间状语,再描述返回舱分离,最后描述返回舱降落,遵循事物发展的一般顺序。

⑤ 将"like one at Kyoto University in Japan"加括号作为补充成分并紧跟在其修饰的词语"telescope"之后。

⑥ 注意在翻译时要体现时态,此处的"spotted"是过去式,因此应译为"曾看见"。

⑦ "a bit of a rush"原意为"匆忙",此处结合上下文语境译为"时间紧张"。

⑧ "The team"作为主语应具体说明指代对象,即"回收小组"。

⑨ "mission manager"的翻译应注意与上下文保持一致,即"任务负责人"。

⑩ 本句中的"brought"为过去式,而谓语动词"hope"为现在式,说明动作发生在过去,即"已经带回"。

⑪ "left"是常用的动词,原意为"离开",但是由于此处的动作主体为"airplane",需将动作具体化且符合动作主体的行为,因此译为"起飞"更加通顺。

⑫ 本句运用了一系列动词,在翻译时应格外注意动作发生的先后顺序,增添连接词使文章更加连贯。

翻译练习 4.10

SpaceX Inspiration4 Updates: The Crew Spends Their First Day in Orbit

1. After a successful launch① on Wednesday, the first space crew with no professional astronauts aboard is circling Earth.

2. The crew of the Inspiration4 mission② is flying safely around Earth, SpaceX said in a Twitter update Wednesday afternoon. But the altitude that they are orbiting at is not as unusual as some space commentators, including me, have said.

3. Andy Tran, a production supervisor for SpaceX,③ spoke during a livestream Tuesday covering the launch of the Inspiration4 mission: "They're going to be higher than the International Space Station, higher than the Hubble Telescope, honestly higher than any humans other than those who went to the moon."

4. That is not true, and I goofed in including that erroneous information in my stories yesterday. A former Times editor④ of mine often talked of facts that were "too good to be true" and this was one of those.

5. The Inspiration4 crew is as far as 366 miles above Earth, which is more than 100 miles higher than the International Space Station. And they are farther from the planet than most astronauts who have gone to space since the end ofNASA's Apollo program in the 1970s.⑤

6. But the space shuttle⑥ mission that deployed the Hubble Space Telescope in 1990 was in an elliptical orbit that went as high as 386 miles above Earth. And in 1999, a mission to repair and upgrade that telescope was in an orbit that reached an altitude of 378 miles. Astronauts on both missions traveled farther from Earth than the crew members of Inspiration4.

7. All of this illustrates how close humans have remained to home since the end of Apollo. The moon is almost 240000 miles from Earth. Since Apollo 17 returned in 1972, no one has traveled more than 400 miles away from the planet, and that will not change until the first crewed mission of NASA's Artemis program, which is tentatively scheduled for late 2023.

8. Unlike with NASA missions, there is little information about the Inspiration4 crew. SpaceX reported on Twitter that the crew members are "healthy, happy and resting comfortably," that they had performed the first round of scientific research, and that they had eaten a couple of meals and slept.

9. SpaceX also tweeted a photograph of the glass dome at the top of the Crew Dragon spacecraft,⑦ although there was no one inside it at the time.

10. As private space travelers and not NASA employees, the SpaceX crew can choose to maintain a veil of privacy around their activities. More images and video will eventually be shown in the final episode of a Netflix documentary series about the mission. It is also possible that the crew could participate in live public broadcasts from space, but no plans have been announced yet.

11. — Kenneth Chang

12. More people are in space right now than ever before.

13. Outer space got a little more crowded on Wednesday night.

14. The four-person crew of SpaceX's Inspiration4 raised the number of people in space to 14, edging out⑧ a record set in 2009 when 13 people lived on the International Space Station after the space shuttle Endeavour docked there.

15. This year, though, the 14 humans in space are on three separate missions.

16. Shenzhou-12, from China, is completing a 90-day trip and is carrying three astronauts. And Expedition 65, from NASA, launched in April with a team of seven, who are currently at the space station.

17. The "Commander & Benefactor" of the Inspiration4 is Jared Isaacman, a high school dropout who became a billionaire founder of a payments processing company.⑨ He follows fellow billionaires Richard Branson, the entrepreneur behind the Virgin companies, and Jeff Bezos, the founder of Amazon, who went this year.

18. Billionaires like them, and the private companies they fund, have made the cost of space travel cheaper, according to Dr. Elliott Bryner, an assistant professor of mechanical engineering at Embry-Riddle Aeronautical University in Arizona⑩. As those costs go down, the number of people who are in space will go up, he said.

19. "The thing that has been barring us from going to space is cost," Dr. Bryner said on Wednesday night. "With private launches, the number of people who can go to space will continue to increase."

20. "It's still a millionaire's game, but at least you don't have to be a superpower country," he said.

21. — Azi Paybarah

22. The crew will spend 3 days aboard a capsule slightly bigger than a minivan.

23. The Crew Dragon is a gumdrop-shaped capsule—an upgraded version of SpaceX's original Dragon capsule,⑪ which has been used many times to carry cargo. It is roughly comparable in size to the Apollo capsule that took NASA astronauts to the moon in the 1960s and '70s. Earlier NASA capsules—Mercury and Gemini—were considerably smaller.

24. The capsule has more interior space than a minivan, but less than a studio apartment. ⑫ And there is a bathroom. As you can probably imagine, you and some of your friends may be able to pile into a space like that for a brief time, but much longer could become uncomfortable.

25. So far, NASA's missions in Crew Dragon have spent no more than about a day orbiting the planet before docking with the space station. Inspiration4's crew will spend three days aboard.

26. "It's like an extended camping trip," Mr. Sembroski said during Tuesday's news conference. "You're in a camper van with some of your closest friends for three days."

27. The crew members will be able to pull out sleeping bags"and strap yourself in so you don't float into each other during the middle of the night," he said.

28. "There will be a couple unique challenges maintaining privacy here and there," Mr. Sembroski said. He said they had received good tips from NASA astronauts who previously traveled to space in the capsule. ⑬

29. "We'll let you know more about how successful they were when we come back," Mr. Sembroski said.

30. While food for spaceflight has made great advancements in quality since the 1960s, dining may not be a highlight of this orbital trip. In the Netflix documentary about Inspiration4, Ms. Arceneaux said during a taste test that she didn't think she'd eat much in space. SpaceX has also not said who prepared the meals for this mission.

31. One of the planned meals is cold pizza, Mr. Sembroski revealed during an episode of an Axios podcast that followed their training for the mission. ⑭

32. "The cold pizza better be packed, because that was my order," Dr. Proctor said on Tuesday. "Food and mood is so important. So I think for us it was really important working with SpaceX to get food that made us feel comfortable."⑮

【参考译文】

美国太空探索技术公司"灵感4号"飞船快报：机组人员完成第一天轨道飞行

1. 周三，随着飞船成功发射升空，第一个平民机组正环绕地球飞行。

2. 周三下午，美国太空探索技术公司在推特上发布最新消息说，"灵感4号"任务的机组人员正在环绕地球飞行，一切安好。我和其他一些航天评论员曾认为这次飞行的轨道高度非同寻常，实际并非如此。

3. 周二，美国太空探索技术公司的生产主管安迪·特兰在"灵感4号"任务发射的线上直播中说，"机组人员的飞行高度会超过国际空间站，也会超过哈勃望远镜。可以说，除了那些曾登月的航天员，没有人到达过他们这次的高度。"

4. 这一说法其实并不准确，而我在昨天的报道中误用了这一错误信息。我的一位《纽约时报》前编辑同事常说，有些情形"好得太离谱"，现在这个报道就是一个例子。

5. "灵感4号"机组人员距地面高度为366英里（585.6千米），比国际空间站高100多英里（160多千米）。自20世纪70年代美国国家航空航天局阿波罗计划结束后，大多数进入太空的航天员达到的高度都逊色于它们。

6. 1990年，安置哈勃空间望远镜的航天飞机在一个椭圆轨道上进行，距离地面高达386英里（617.6千米）。1999年，维修和升级哈勃望远镜的航天器的行进轨道距地球378英里（604.8千米）。完成这两次任务的航天员距地面高度都超过"灵感4号"机组成员。

7. 由此可见，自阿波罗计划结束以来，人类未曾远离过地球。月球距离地球几乎24万英里（38万千米）。自1972年"阿波罗17号"返回以来，人类的航天活动一直在距地球400英里（640千米）以内的范围进行。这种情况将会一直持续到美国国家航空航天局阿尔忒弥斯计划的第一次载人任务才会有所改变。该任务暂定2023年年底进行。

8. 与美国国家航空航天局所执行的任务不同的是，外界对"灵感4号"机组人员的情况知之甚少。美国太空探索技术公司发推文称，机组成员目前"身体状况良好、心情不错、得到充分休息"，推特还称他们已经完成了第一轮科学研究，吃了几顿饭，也睡了觉。

9. 美国太空探索技术公司还在推特上发布了一张照片，显示的是载人"龙"飞船顶部的玻璃穹顶，不过当时那里并没有人。

10. 这次的机组人员都是平民太空旅行者，并非美国国家航空航天局雇员，所以可以选择不对外透露自己的活动情况。网飞公司负责制作这次任务的系列纪录片。以后，观众可以在纪录片的结局部分看到更多相关影像。机组成员也许会在太空参加公开直播，但目前相关计划尚未公布。

11. ——肯尼斯·张

12. 太空探索,盛况空前。

13. 周三晚的外太空格外热闹。

14. 随着美国太空探索技术公司"灵感4号"任务的四人机组进入太空,目前身处太空的人数达到14人,刚好打破2009年创下的纪录。当时"奋进号"航天飞机与国际空间站对接后,空间站内有13人。

15. 但今年身处太空的14人分别执行三项不同的任务。

16. 搭载了三名航天员的"神舟十二号"来自中国,正在执行一项为期90天的任务。于四月发射的美国国家航空航天局"远征65号"搭载的七名机组人员目前身处空间站。

17. 贾里德·艾萨克曼是"灵感4号"飞行船的"指挥官和赞助人"。虽然高中都未毕业,但因创办了一家支付处理公司而成为亿万富翁。他是继维珍集团创始人理查德·布兰森和亚马逊创始人杰夫·贝索斯后,今年又一名飞往太空的亿万富翁。

18. 艾略特·布里纳博士是美国亚利桑那州安柏瑞德航空大学的机械工程助理教授。据他说,他们这样的亿万富翁及其资助的私营公司降低了太空旅行的成本。随着成本下降,进入太空的人数将增长。

19. 布里纳博士周三晚上表示,"一直以来我们只能望'空'兴叹,就是因为成本太高。随着搭载平民的飞船成功发射,进入太空的人数会不断增加。"

20. 他说,"虽然太空旅行仍只是百万富翁的消遣游戏,但至少不是只有超级大国才能做到了。"

21. ——阿齐·帕伊巴拉

22. 航天员将在一个略大于小型货车的太空舱度过三天。

23. 载人"龙"飞船的太空舱形似橡皮糖,由美国太空探索技术公司最初的"龙"飞船太空舱升级而来,而最初的"龙"飞船曾多次用于运送货物。载人"龙"号飞船大致上与阿波罗计划的太空舱大小相当。20世纪六七十年代,阿波罗太空舱曾运送美国国家航空航天局航天员登月。相比之下,更早的美国国家航空航天局水星和双子星计划使用的太空舱要比载人"龙"飞船小很多。

24. 载人"龙"飞船太空舱的内部空间比小型货车宽敞,但比单间公寓狭小,内设有一个卫生间。可想而知,在如此狭小的空间,你和几个朋友一时半会儿也许能将就,但很快就会难以忍受。

25. 截至目前,美国国家航空航天局载人"龙"飞船执行的任务都是在绕地球仅一天左右就与空间站对接了。而"灵感4号"机组人员将在飞船上待三天。

26. "这就像一次郊游露营的加长版,"森布罗斯基先生在周二的新闻发布会上说,"你得和几个密友在露营车里待三天。"

27. 他还说,机组成员届时会拉出睡袋,"把自己绑在里面,这样半夜里就不会随处飘荡,相互碰撞了。"

28. 森布罗斯基先生说,"总有一些意想不到的状况出现,使得机组人员隐私难保。"他说,美国国家航空航天局宇航员此前曾搭乘此类太空舱进入太空,他们就此提供了很好的建议。

29. 森布罗斯基先生说,"等我们回来再告诉你们这些建议管不管用。"

30. 20世纪60年代以来,航天食品的质量已有很大提升,但用餐不一定是这次轨道旅行的亮点。在网飞公司制作的"灵感4号"任务纪录片中,阿尔克诺女士试吃时说,她觉得自己在太空不会吃太多东西。美国太空探索技术公司也没有透露这次任务的食品供应方。

31. 爱可信播客记录了机组四人为完成此次任务而接受训练的情况。森布罗斯基先生在其中一集中透露,备好的食物中有一道是冷披萨。

32. "冷披萨最好都已经打包好了,因为我喜欢那样,"普罗克特博士周二说,"食物和心情非常重要。所以我觉得对我们来说,和美国太空探索技术公司一起准备好可口的食物真的很重要。"

【注释】

① 在转换成汉语时,需补充出"launch"的主语,这样句子意思更清晰。

② "mission"在航空航天科技英语中通常指太空探测任务,如火星探测、月球探测等,也特指执行探测任务的航天器,如此处的"灵感4号"飞船。

③ 当英语中同位语为主语的补充描述且较短时,可将其译为汉语主语的前置定语,这样可以避免另起小句,导致句子重心转移。

④ 汉语在名词后多用范畴词,此处作者身份为记者,给"editor"一词补充说明,加上同事身份,会让句意更清晰。

⑤ 对于较长的复合句,翻译时需要调整语序,例如此句中,可以将"自阿波罗计划结束后"这一时间状语前置,同时将主语转换为原句介词"than"的宾语。

⑥ "space shuttle":航天飞机,是一种载人往返于近地轨道和地面间的有人驾驶、可重复使用的运载工具;既能像运载火箭那样垂直起飞,又能像飞机那样在返回大气层后在机场着陆;由轨道器、外贮箱和固体助推器组成。航天飞机最早由美国研发,是美国国家航空航天局1981—2011年运作的近地轨道航天器,也是美国航天飞机计划的核心。美国共制作五架完整的航天飞机轨道飞行器并在前后三十余年间执行135次任务,每次都是从佛罗里达州的肯尼迪航天中心起飞。除发射哈勃空间望远镜、多枚人造卫星和空间探测器外,这些任务还完成轨道科学实验,参与国际空间站的建设和维护。著名的航天飞机有美国的哥伦比亚号(Columbia)、挑战者号(Challenger)、发现号(Discovery)、亚特兰蒂斯号(Atlantis)和奋进号(Endeavour),以及苏联的暴风雪号。

⑦ 翻译中可适当拆分较长的后置定语,如本句先说发布了一张照片,再以小句描述照片内容,避免"照片"前的定语过长。

⑧ "edge out":该短语意为以微弱数量超过或取胜。

⑨ 根据逻辑关系可适当调整语序,直译为"亿万富翁创始人"表意不明,处理为"因为创办公司而成为亿万富翁"更符合逻辑。

⑩ 当同位语较长时,应避免将其简单处理成主语的前置定语,因为这样会导致定语过长,不符合汉语的表达习惯,可将其分译。具体到此处,可以处理成判断句,单独成句:艾略特·布里纳博士是美国亚利桑那州安柏瑞德航空大学的机械工程助理教授。

⑪ 将名词词组动词化处理,若保留英文结构译为"……的升级版"则定语过长,此处可以将其转换为动词"升级"。

⑫ 英语同一单词译入汉语时,需要根据语境,灵活转换,例如此句中"more"和"less"若直译为"更多"和"更少"则略显生硬,可以根据语境,与空间搭配,译为"宽敞"和"狭小"更合理。

⑬ 修饰 NASA 航天员的定语过长,可适当分译。主语"他们"与介词宾语"NASA 航天员"在句中的位置可相互转换,使得汉语小句主语保持一致。

⑭ 翻译长句时可将句子各成分分译并调整顺序,避免受英语原句结构的影响。此句主语为"Mr. Sembroski",谓语为"revealed",宾语为前置的小句"One of the planned meals is cold pizza"说明 revealed 的内容,"during an episode of an Axios podcast"为状语说明 revealed 的时间,"that followed their training for the mission"为定语从句修饰 podcast。翻译时将定语从句分译并位置提前。

⑮ 对于以"it"作形式主语的句子,翻译时可先说明"it"指代的情况,再补充"it"的表语。

翻译练习 4.11

China's Mars Rover Mission Lands on the Red Planet

1. The success① establishes China as a principal contender in what some see as a new era of space competition.

2. The United States now has a company② on Mars.

3. A Chinese spacecraft descended through the thin Martian atmosphere and landed safely on a large plain on Saturday morning, state media reported, accomplishing a feat that only two other nations had before.③④ (In the United States, it was still Friday—7:18 p. m. Eastern time—when the spacecraft touched down.)

4. The landing follows China's launch last month of the core module of a new orbiting space station, as well as a successful mission in December that collected nearly four pounds of rocks and soil from the moon and brought it to Earth. Next month, the country plans to send three astronauts into space, inaugurating what

could become a regular Chinese presence in Earth's orbit.

5. Just by arriving at Mars and orbiting it in February, China's space program confirmed its place among the top tier of agencies exploring the solar system[⑤]. Now that it has executed a landing—with the deployment of a rover still to come— it has established itself as a principal contender in what some view as a new era of space competition.

6. "China's successful Mars landing demonstrates to the world that there is another country with advanced interplanetary space capacities," said Namrata Goswami, an independent analyst and co-author of a new book on space exploration, "Scramble for the Skies."

7. Thomas Zurbuchen, NASA's associate administrator for science[⑥], offered his congratulations to China. "Together with the global science community, I look forward to the important contributions this mission will make to humanity's understanding of the Red Planet," he wrote on Twitter.

8. Until Friday, the China National Space Administration had said little about its plans for the landing, in keeping with its usual secrecy involving operations. The news of the impending landing, however, began to spill out[⑦] on social media and in official news reports, signaling that it was imminent.

9. China's state television did not report on the landing live, but later devoted hours of programs to the mission.

What is China's mission to Mars?

10. TheTianwen-1 mission launched from Earth last July, aiming to take advantage of the window of time every two years when Mars and Earth are closest together during their voyages around the sun.

11. The mission consists of an orbiter, a lander and a rover.

12. TheTianwen-1 orbiter pulled into Martian orbit on Feb. 10; since then, it has been circling at a safe distance, preparing for the landing attempt.

13. The unnamed landing craft carries a rover, which was named Zhurong, after a god of fire in Chinese folk tales. That name beat out ⑧ nine other semifinalists that were announced in February.

14. The mass of Zhurong is about 240 kilograms, or roughly 530 pounds. That is a bit heftier than the Spirit and Opportunity rovers that NASA landed on Mars in 2004, but only about one-fourth the mass of the two currently operating NASA Mars rovers, Curiosity and Perseverance.

15. Days after the touchdown, the rover will roll off the lander. Like Spirit and Opportunity, Zhurong will be powered by solar panels, which are retractable

so that it can periodically shake off any accumulated dust. For Perseverance and Curiosity, nuclear batteries turn heat released by the decay of radioactive plutonium into electricity.

16. Zhurong's seven instruments include cameras, ground-penetrating radar, a magnetic field detector and a weather station.

17. This was not China's first attempt at a Mars mission. That was Yinghuo-1, which failed nearly 10 years ago, through no fault of the country's own. That spacecraft burned up in Earth's atmosphere when the Russian rocket that was carrying it failed in flight⑨.

Where did the rover land and what will it study?

18. It landed in Utopia Planitia, or Nowhere Land Plain, a huge basin a couple of thousand miles wide in the northern hemisphere that was most likely carved out by a meteor impact. The same region was visited by NASA's Viking 2 lander in 1976.

19. The plains are part of the northern lowlands of Mars. If there was once bountiful water on the red planet a few billion years ago, this region could have been underwater, part of an ocean covering the upper part of the planet. Utopia Planitia lies lower than features that have been proposed as two sets of shorelines, remnants from such early Martian oceans.

20. Some of the water from that hypothesized ocean may once have percolated underground and could still be frozen there today. In 2016, scientists using a radar instrument on NASA's Mars Reconnaissance Orbiter concluded that there is indeed much ice there—as much water as Lake Superior spread over an area bigger than New Mexico.

21. One goal of the Tianwen-1 mission is to better understand the distribution of ice in the region, which future human colonists on Mars could use to sustain themselves.

22. The Chinese space agency has highlighted the international collaboration on the Tianwen-1 mission, including contributions from the European Space Agency, Argentina, France and Austria.

How did the rover land on Mars?

23. Landing on the red planet is perilous—NASA engineers refer to it as seven minutes of terror when its rovers, most recently Perseverance, arrive.

24. Because Tianwen-1 was already in orbit around Mars, its incoming speed was not quite as fast as Perseverance's. Thus, China's lander required a bit of extra terror—nine minutes—for the landing, Global Times, a newspaper reported on

Friday, citing experts. The probe was also operating on its own, as signals currently take 17 minutes 42 seconds to travel between Mars and Earth.

25. Spacecraft descend toward Mars at a high speed, and the thin atmosphere does not do enough to slow the trip to the ground. The shock waves of air compressed by the speeding capsule generate extreme heat that must be absorbed or dissipated. A number of Soviet, and European missions have crashed.

26. Only NASA has reached the surface of Mars intact more than once. The landings of its largestrovers, Curiosity and Perseverance, have relied on parachutes to slow the spacecraft, shields to dissipate the heat from atmospheric friction and intricate systems called sky cranes. These were basically rocket-powered jetpacks, which carried the rovers beneath them and lowered them to the surface on cables before flying safely away from the landing zone.

27. "For our country's first Mars exploration mission, we didn't have firsthand data about the environment on Mars, especially the atmosphere," Chen Baichao, a senior designer for the mission, said in remarks reported by The Paper, a Shanghai-based news site. "So it was tantamount to us entering a completely unknown environment, and you can imagine how difficult that is."

28. Global Times reported that theTianwen-1 probe lowered its altitude from its parking orbit before its lander-rover combination separated with the orbiter at around 4 p.m. Friday, Eastern time. (In China, it was 4 a.m. Saturday.)

29. The orbiter then rose and returned to its parking orbit about half an hour after the separation, to provide relay communication for the landing craft combo, the Chinese space agency told Global Times. The lander-rover combination circled Mars for another three hours before entering the Mars atmosphere en route to landing.

What else has China's space program been doing?

30. For the Tianwen mission, a cone-shaped entry capsule carried the lander and rover through the atmosphere. A heat shield protected the spacecraft from superheated gases as it sped through the top of the atmosphere. Then the friction of the thin Martian air helped it slow down—by about 90 percent, Tan Zhiyun, a designer at the China Academy of Space Technology, told Global Times.

31. At a lower altitude, the heat shield was jettisoned. At the next step, the parachute and the top nose-shaped piece were discarded. Firing a rocket engine, the four-legged lander, similar in design to the Chang'e-3 and Chang'e-4 lunar landers, then hovered briefly as it searched for a safe spot and descended toward a safe powered landing.

32. China is on the far side of the moon with a robotic rover, and it recently put a large piece of its next space station in Earth orbit. But its space program has many other goals, too.

What else has been happening on Mars recently?

33. In recent weeks, the surface of Mars has seen its first flights by Ingenuity, an experimental helicopter built by NASA. The small rotorcraft was meant to demonstrate that it was possible for something to fly like a helicopter or airplane in the thin Martian atmosphere.

34. So far, the mission has been a great success. The first flight, on April 19, hovered at an altitude of some 10 feet for about 30 seconds. Subsequent attempts have been more ambitious, with the fifth flight making a one-way trip across 423 feet at a higher altitude. From the new location, Ingenuity will begin a secondary mission of serving as something like an aerial scout for's Perseverance rover.

35. NASA and China were actually beat ^{⑩} to Mars this year by Hope, an orbiter built by the United Arab Emirates. It has been circling the red planet since Feb. 9 and has begun making scientific observations of the planet's atmosphere and weather.

36. Earlier this week, the spacecraft released images of hydrogen atoms surrounding Mars at different times of day. Such studies may help scientists on Earth better understand how gases circulate on Mars and eventually escape^{⑪} the planet.

【参考译文】

中国火星探测器登陆火星

1. 中国成功登陆火星意味着其在所谓的太空竞争新时代中头号竞争者的地位已确立。

2. 如今美国在火星上不再形单影只了。

3. 据中国官方媒体报道,随着中国探测器于周六上午穿过稀薄的火星大气层,安全降落在一片大平原上,中国成为第三个完成此项壮举的国家。(探测器着陆时间为美东时间星期五晚上7点18分。)

4. 在此之前,中国于上月发射了一个新的轨道空间站核心舱,去年12月还成功地从月球采集了近4磅(1.81千克)的岩块和土壤带回地球。下个月,中国计划将三位航天员送入太空,至此中国将成为地球轨道上的常客。

5. 今年2月,中国航天器登陆火星并环绕火星飞行。仅凭此举,在全球探索太阳系的第一梯队中,中国航天计划的地位得到巩固。目前,中国已完成一次着陆,随后还会有一辆火星车进行探测活动。这样,在所谓的太空竞争新时代,中国已完全确立了其头号竞争者的地位。

6. "中国成功登陆火星向世界证明,除了美国以外,中国也拥有先进的星际空间能力,"纳姆拉塔·哥斯瓦米说。他是独立分析师及最新出版的太空探索书籍《角逐苍穹》的合著者。

7. 美国国家航空航天局科学署副主管托马斯·泽布臣/科学任务理事会副主任托马斯·泽布臣向中国表示祝贺。他发推文说:"我与全球科学界共同期待,中国的这次任务能为人类了解这颗红色星球做出重要贡献。"

8. 中国国家航天局沿袭以往的保密传统,对本次着陆计划几乎避而不谈,直到本周五,社交媒体和官方报道开始透露探测器即将着陆,着陆就在眼前了的消息开始不胫而走。

9. 虽然中国国家电视台没有直播这次着陆,却在之后的节目中用了几个小时介绍这次任务。

中国登陆火星,目的何在?

10. "天问一号"发射升空是在去年7月,目的是利用火星和地球在绕太阳飞行时每两年距离最近的窗口期。

11. 探测器由轨道飞行器、着陆器及月球车三部分组成。

12. 2月10日,"天问一号"探测器进入火星轨道,此后一直在安全高度绕火星运行,为着陆做准备。

13. 未命名的着陆器搭载了一辆名为"祝融号"的火星车,因中国民间传说中的火神祝融而得名。在2月份公布的十个入围名单中,这个名字最终脱颖而出。

14. "祝融号"质量约为240千克,略重于美国国家航空航天局2004年登陆火星的"勇气号"和"机遇号"探测器,但只有现役"好奇号"和"毅力号"探测器质量的四分之一左右。

15. 在着陆后的几天内,"祝融号"会驶出着陆器。与"勇气号"和"机遇号"一样,"祝融号"靠太阳能电池板供电。这种电池板可伸缩,有利于探测器定期抖落积尘。而"毅力号"和"好奇号"则靠核电池供电。这种电池能将放射性钚衰变释放的热量转化为电能。

16. "祝融号"上配备有7种仪器,包括各种照相机、一台探地雷达、一台磁场探测器和一台气象测量仪等。

17. 这并不是中国首次探测火星。大概10年前的"萤火一号"才是!那时"萤火一号"发射失败,但问题不在中国,而是当时搭载"萤火一号"的俄罗斯火箭发射失败,在地球大气层中烧毁所致。

火星车在哪里着陆,其将进行哪些研究?

18. 着陆器降落在乌托邦平原,或称天堂平原,这是火星北半球的一个几千英里宽的巨大盆地,极有可能是由陨石撞击而成。1976年,美国国家航空航天局的"海盗

2 号"着陆器也曾造访过此地。

19. 这些平原是火星北部低地的一部分。如果几十亿年前这颗红色星球上曾经存在丰富的水资源,那么当时这片区域很有可能位于水下,是覆盖火星表面的某海洋的一部分,因为乌托邦平原的地势要低于两段海岸线,而这些海岸线据称是早期火星海洋遗迹。

20. 这个假定海洋中的一些海水可能曾渗透到地下,至今也许仍在那里处于冻结状态。2016 年,科学家们利用美国国家航空航天局的火星勘测轨道飞行器上的雷达仪器,推断火星上目前确实存在大量冰,其体量相当于面积比美国新墨西哥州还大的苏必利尔湖的水量。

21. "天问一号"的一项任务就是进一步了解该地区冰的分布情况,便于未来移居火星的人类维持生活使用。

22. 国家航天局强调国际协作在"天问一号"探测任务中发挥了重要作用,欧洲航天局、阿根廷、法国和奥地利都做出了贡献。

火星车如何着陆火星

23. 在红色星球上着陆相当危险。最近一次,"毅力号"抵达火星时,美国国家航空航天局的工程师称之为"恐怖七分钟"。

24. 因为"天问一号"已进入火星轨道运行,所以它进入火星大气层的速度不像"毅力号"那样快。因此,中国《环球时报》周五援引专家评论称,中国着陆器需要经历更长的恐怖时间,即九分钟,才能着陆。"天问一号"上的探测器也是自主运行,因为目前信号在火星和地球之间传播需要耗时 17 分 42 秒。

25. 着陆器以高速着陆火星,而火星稀薄的大气层不足以减缓其冲向地面的速度。高速飞行的太空舱压缩生成的空气冲击波会产生高温,而这个高温必须想方设法被吸收或消解掉。美国、苏联和欧洲就曾因为这个原因导致多次任务失败。

26. 只有美国国家航空航天局曾不止一次顺利到达火星表面。美国最大的火星车"好奇号"和"毅力号"着陆时,用降落伞来减缓飞船的速度,用散热盾来减少由大气摩擦和被称为"空中起重机"的复杂系统所产生的热量。这些系统实际上就是喷气背包,由火箭驱动,下面载有火星车,并在其安全飞离着陆区之前,用缆绳将火星车放至地面。

27. "由于这是我国首次火星探测任务,对类似火星环境,特别是大气这样的参数,我们没有第一手数据",此次任务的高级设计师陈百超在接受上海新闻网站澎湃新闻的报道时表示,"对我们来说,相当于到了一个完全未知的环境,难度可想而知。"

28. 据《环球时报》报道,"天问一号"探测器从其中继轨道降低高度,随后于美国东部时间周五下午 4 点左右(中国时间周六凌晨 4 点),其着陆器-火星车组合体与轨道飞行器分离。

29. 《环球时报》来自国家航天局的消息称,分离大约半小时后,轨道飞行器上升并返回中继轨道,为着陆器-火星车组合体提供中继通信。在进入火星大气层准备着陆的过程中,组合体在火星上空又盘旋了 3 小时。

中国的太空计划还包括什么?

30.“天问一号”的锥形进入舱携带着陆器-火星车组合体穿过火星大气层。进入舱高速穿过大气层顶部时,散热盾保护其免受过热气体的影响。中国空间技术研究院的设计师谭志云在接受《环球时报》采访时表示,在与火星的稀薄空气摩擦过程中,进入舱的速度降低了约 90%。

31. 在降至一定高度时,隔热盾抛离。之后,降落伞和顶部鼻形部件也一同被抛离。随即,火箭引擎启动。这个与“嫦娥三号”和“嫦娥四号”月球着陆器设计相似的四脚着陆器先是在空中短暂盘旋以寻找安全着陆点,然后安全着陆。

32. 目前中国在月球背面有一辆机器月球车,近期已经将未来空间站所需的大部件送入地球轨道。此外,中国的太空计划还包括许多其他目标。

近期火星上还发生了什么?

33. 最近几周,美国国家航空航天局制造的实验直升机“机智号”在火星表面首飞成功。设计试飞这架小型旋翼机就是为了展示,直升机或飞机之类的飞行器可以在火星稀薄的大气中飞行。

34. 迄今为止,这项任务已取得了巨大成功。首飞是在 4 月 19 日,“机智号”在高度大约 10 英尺(3.04 米)的空中盘旋了 30 秒左右。之后几次试飞高度逐渐增加,第五次就实现了在 423 英尺(128.9 米)高度进行单程飞行。此外,“机智号”将在新地点开启第二项任务,为美国国家航空航天局的“毅力号”探测车提供空中侦察。

35. 今年,阿拉伯联合酋长国建造的轨道飞行器(火星探测器)“希望号”实际上已经让美国和中国的火星计划甘拜下风。自 2 月 9 日以来,“希望号”一直在绕火星运行,并已经开始对火星的大气和天气状况进行科学观测。

36. 本周早些时候,“希望号”发布了火星周围的氢原子在一天中不同时间的图像。通过这些研究,科学家可以更好地了解气体在火星上是如何循环流动并最终逃逸的。

【注释】

① “The success”应根据上下文补充信息,翻译成“中国成功登陆火星”,这样文意更加通顺,便于读者理解。

② “has a company”可使用四字格和正反译法,译为“不再形单影只了”,使表达更加生动。

③ “a feat that only two other nations had before”若直译为“这是以前只有另外两个国家拥有的壮举”则表达生硬,可以灵活处理成“中国成为第三个完成此项壮举的国家”。

④ 此句在表达上应首先梳理句子的逻辑关系,分清层次和主次,再补充“随着”等逻辑词突出句内关系。

⑤ 此句应首先拆分原句的结构,先译“among the top tier of agencies exploring

the solar system",再译句子重点"China's space program confirmed its place",避免冗长句式。此处可将主动句译为被动句,译为"中国航天计划的地位得到巩固",更符合汉语表达习惯。

⑥ 此处应通过查阅资料确认人物背景信息,并注明出处,以保证翻译的准确性。

⑦ "spill out"指液体从容器中溢出,此处根据上下文译为四字格"不胫而走",表达更加生动。

⑧ "beat out"原意为"击败",此处使用四字格"脱颖而出",准确生动,符合汉语表达习惯。

⑨ 此句首先应梳理句子逻辑,再按照因果顺序进行翻译,应先翻译从句"when the Russian rocket that was carrying it failed in flight",再翻译主句"that spacecraft burned up in Earth's atmosphere"。

⑩ "beat"在被动语态中原意为"被打败",此处使用四字格"甘拜下风",形象准确地表达了原文含义。

⑪ "escape"原意为"逃脱""逃避",在航空航天科技英语中指气体从行星表面"逃逸"。

翻译练习 4.12

50 Years Ago, NASA Put a Car on the Moon

1. The lunar rovers of Apollo 15, 16 and 17 parked American automotive culture on the lunar surface, and expanded the scientific range of the missions' astronaut explorers.

2. Dave Scott was not about to pass by an interesting rock without stopping. ① It was July 31, 1971, and he and Jim Irwin, his fellow Apollo 15 astronaut, were the first people② to drive on the moon. After a 6-hour inaugural jaunt in the new lunar rover, the two were heading back to their lander, the Falcon, when Mr. Scott made an unscheduled pit stop. ③

3. West of a crater④ called Rhysling, Mr. Scott scrambled out of the rover and quickly picked up a black lava rock, full of holes formed by escaping gas. Mr. Scott and Mr. Irwin had been trained⑤ in geology and knew the specimen, a vesicular rock, would be valuable to scientists on Earth. They also knew that if they asked for permission to stop and get it, clock-watching mission managers would say no. So Mr. Scott made up a story that they stopped the rover because he was fidgeting with his seatbelt. The sample was discovered when the astronauts returned to Earth, Mr. Scott described what he'd done, and "Seatbelt Rock" became one of the most prized geologic finds from Apollo 15. ⑥

4. Like many lunar samples returned to Earth by the final Apollo missions, Seatbelt Rock never would have been collected if the astronauts had not brought a car with them. Apollo 11 and Apollo 13 are the NASA[7] lunar missions that tend to be remembered most vividly. But at the 50th anniversary of Apollo 15, which launched on July 26, 1971, some space enthusiasts, historians and authors are giving the lunar rover its due as one of the most enduring symbols of the American moonexploration program.

5. Foldable, durable, battery-powered and built by Boeing and General Motors[8], the vehicle is seen by some as making the last three missions into the crowning achievement of the Apollo era.

6. "Every mission in the crewed space program, dating back to Alan Shepard's first flight, had been laying the groundwork for the last three Apollo missions," said Earl Swift, author of a new book about the lunar rover, "Across the Airless Wilds: The Lunar Rover and the Triumph of the Final Moon Landings."

7. "You see NASA take all of that collected wisdom[9], gleaned over the previous decade in space, and apply it," Mr. Swift said. "It's a much more swashbuckling kind of science."

8. Once Neil Armstrong's small step satisfied Project Apollo's geopolitical goals, NASA emphasized science, said Teasel Muir-Harmony, curator of the Apollo collections at the Smithsonian Institution's National Air and Space Museum in Washington. While the first moon-walkers retrieved samples near their landing sites, scientists had long hoped for a lunar road trip that promised rare rocks. Plans for a lunar rover were finally given the green light just two months before Armstrong and Buzz Aldrin became the first humans on the moon.

9. Though moon buggies had been imagined for years, driving a car on the moon is more complicated than it sounds. Throughout the 1960s, engineers studied a variety of concepts: tank-like tracked vehicles, flying cars, even a rotund monstrosity shaped, as Mr. Swift describes it, "like an overgrown Tootsie Pop, with its spherical cabin up top of a single long leg, which in turn was mounted on a caterpillar-tread foot." Ultimately, a car-like buggy came out on top.

10. "There were other outlandish ideas, like a pogo stick, or a motorcycle—things that I am glad they didn't pursue,"[10] Dr. Muir-Harmony said. "The lunar rover is, in some ways, relatively practical."

11. The moon car was also quintessentially American. The rover's exposed chassis, umbrella-like antenna and wire wheels meant it looked like no car on Earth, yet its connection to the American auto industry and the nation's love affair

with the automobile captivated public attention like nothing since Apollo 11, Dr. Muir-Harmony said.

12. Starting with Project Mercury in the 1960s, a Florida car dealer allowed astronauts to lease Chevrolet cars for $1, which were later sold to the public. The Apollo 15 crew chose red, white and blue Corvettes. A photo spread in Life magazine showed the astronauts posing with their iconicAmerican muscle cars alongside the moon buggy, making the lunar rover look cool by association, Dr. Muir-Harmony said. "There's a lot to unpack in that picture," she added.

13. Mr. Irwin and Mr. Scott helped drum up excitement[①] once they and the rover reached the moon. During the mission's second day, the astronauts drove to a crater named Spur, where they found a large white crystalline rock, a type of mineral on geologists'wish lists[②] because it might provide clues about the moon's origins.

14. The astronauts could barely contain their glee: "Oh, boy!" Mr. Scott shouted. "Look at the glint!" Mr. Irwin said. "Guess what we just found?" Mr. Scott radioed to Earth, as Mr. Irwin laughed with joy. "Guess what we just found! I think we found what we came for."[③]

15. The white rock was later named Genesis Rock,[④] because scientists initially thought it dated to the moon's formation.

16. The astronauts'excitement,[⑤] and their car, brought the Apollo missions back down to Earth,[⑥] Dr. Muir-Harmony said. "It provided a point of access, even as the exploration of the moon was becoming increasingly complex and complicated to follow."

17. Mr. Swift notes that some news reports at the time considered the rover an "inevitable,[⑦] almost comic product of the most automotive people on Earth," although there was nothing inevitable about this extraterrestrial horseless carriage.[⑧]

18. To travel along with the astronauts instead of using a separate rocket, the rover had to weigh less than 500 pounds, but bear twice that in human and geological cargo. On the moon, it had to operate in temperature swings of more than 500 degrees Fahrenheit between sunlight and shade; withstand abrasive lunar dust and micro-meteoroids traveling faster than bullets; and cover a sharp, rugged surface that contained mountains, craters, loose gravel and powder. GM and Boeing engineers scrambled to finish their design in time for the final Apollo missions under threats that NASA would cancel the rover program before it ever left the ground.

19. "If it hadn't been for a couple of engineers at General Motors, there wouldn't have been a rover at all," Mr. Swift said in an interview.

20. His book also explains that immigrant engineers, including Mieczyslaw Gregory Bekker, raisedin Poland, and Ferenc Pavlics, who was born in Hungary, persevered despite large budget overruns, blown deadlines and technical challenges. Though astronauts tend to claim more of the spotlight, engineers played seminal roles in the space program, Mr. Swift said, and some like Mr. Bekker and Mr. Pavlics highlighted the impact that immigrants had on American innovation.

21. "America's race to reach the moon, both within NASA and at the aerospace companies that built the hardware, relied on the minds and talents of immigrants— on Americans who happened to start their lives elsewhere," he wrote.

22. Once the rover arrived and astronauts unfolded it on the moon, the experience of driving was also unexpectedly odd. Astronauts compared it to other Earthly conveyances: Mr. Irwin said the car rose and fell like"a bucking bronco," and Mr. Scott said it fishtailed like a speedboat when he tried to turn at the breakneck speed of 6 miles per hour.

23. Mission managers planned for the rover to travel only as far as the astronauts could walk, in case anything happened and they had to hoof it back to their spacecraft. But Apollo crews covered greater distances with every mission as NASA's confidence grew. When the astronauts left the moon, the rovers were left at the landing sites, where they remain, gathering dust and cosmic rays. Spacecraft orbiting the moon occasionally take their pictures, and in some images, rover tracks are visible.

24. Astronauts found more interesting rocks, enabling scientists to ask different types of questions, said Barbara Cohen, a planetary scientist at NASA's Goddard Space Flight Center in Greenbelt, Md. , who studies the samples. The rover also allowed astronauts to focus on science more than worrying about running out of oxygen or other consumable resources, she said.

25. She recalled participating in a NASA analogue mission several years ago, where scientists would don spacesuits and carry out experiments in a desert field station as though they were on the moon or Mars. She remembered participants getting ready to collect a sample and being interrupted by mission controllers who wanted to check their vitals.

26. "We were like, 'Come on,'" she recalled. "That drove home to me that the geology is not solely in charge. That's one thing the rover does for you; it enables different science questions to be posed that can be more answerable at specific

sites. ”

27. Genesis Rock, a mineral dating to the moon's earliest days, exemplifies Dr. Cohen's point. Scientists are still debating—heatedly—how the moon came to be and what conditions were like there, and by extension, here on Earth, for the first billion years.

28. Dr. Cohen is among several scientists[19] preparing to open untouched samples that have been sealed since they were returned home during the Apollo 17 mission. She will study noble gases in the samples to understand how solar radiation affects moon dust.

29. Katherine Burgess, a geologist at the U. S. Naval Research Laboratory in Washington, D. C. , will study the pristine samples to measure how radiation from the solar wind affects hydrogen and helium levels inside moon dust. Spacecraft can detect helium on the moon from orbit, but scientists still don't know how it varies across lunar terrain. "Without those samples to confirm it, it's still just an open question," she said.

30. Future missions might use lunar helium, especially a variant called helium-3, as a fuel source for nuclear reactors. That means a future generation of lunar rovers may be powered by a material the first generation identified the presence of a half-century ago.

31. Even as scientists study those original samples, many are hoping for a fresh batch, sent home with a new generation of astronauts or collected by rovers descended from the original version. In May, General Motors announced a partnership with Lockheed Martin to build a new rover for NASA's Artemis program, which aims to return American astronauts to the Moon this decade.

32. Although they were built decades apart and by different teams, the lunar rover program informed the first generation of Mars rovers, too, especially Sojourner, the first vehicle on another planet. Engineers at the Jet Propulsion Laboratory, where NASA Mars rovers are built, designed six-wheeled, flexible-framed rovers in a similar vein as early GM designs, Mr. Swift said. "I do think you find an inspirational lineage in that early GM work," he said.

33. Science drives today's NASA more than geopolitics, but the space agency still promotes and carries out human space travel for reasons that go beyond rock prospecting. Dr. Muir-Harmony said the lunar rovers of Apollo, and its modern successors,[20] represent that sense of adventure.

34. "Science is such an important outcome of Apollo, but it is important to recognize what the public is engaged with.[21] The appeal of the lunar rover is

connected to the appeal of human spaceflight, which is being able to witness their joy and a sense of vicarious participation," ② she said.

35. Plus, the adventure of driving across the moon, the greatest road trip of all time, is hard to resist.

36. Then and now, "samples and material from the moon are not getting the focus of public attention," she said. "The rover is."

【参考译文】

50 年前，美国国家航空航天局将汽车送上月球

1. 阿波罗 15 号、16 号和 17 号月球车不但让美国汽车文化驻足月球，还拓展了航天员的科学探索版图。

2. 大卫·斯科特是不会让任何一块稀奇的岩石和自己擦肩而过的。那是 1971 年 7 月 31 日，他和同为"阿波罗 15 号"航天员的詹姆斯·艾尔文首开先河，成为最早驾车穿行于月球表面的人。初次试驾 6 小时后，两人正要返程回到"猎鹰号"登月舱时，斯科特出乎意外地把车停下了。

3. 在雷斯灵环形山西侧，斯科特爬出月球车，迅速捡起一块火山石。这块黑色石头上布满了从火山口泄出的气体形成的孔洞。斯科特和欧文都专门学习过地质学知识，知道这块多孔火山石标本对地球上的科学家来说价值非凡。如果当时申请停车去捡取火山石，肯定会遭到负责计时的任务管理员的拒绝，对此他们心知肚明。于是斯科特找了个借口，说需要停车来调整安全带。等到航天员返回地球后，这块标本才公之于众，斯科特也坦白了这段插曲，而"安全带岩块"也成为"阿波罗 15 号"最珍贵的地质发现之一。

4. 和"阿波罗号"最后几项任务带回的众多月球标本一样，这次如果没有月球车的帮助，航天员是绝对无法收集到"安全带岩块"的。"阿波罗 11 号"和"阿波罗 13 号"也许是美国国家航空航天局最令人难忘的登月任务。在庆祝"阿波罗 15 号"发射 50 周年之际（发射于 1971 年 7 月 26 日），一些航天爱好者、历史学家和作家认为，月球车功劳卓著，算得上美国探月计划最经久不衰的象征之一。

5. 靠电池驱动的月球车既耐用又可折叠，由波音公司和通用汽车公司联合制造。有人认为，阿波罗时代最后三次任务之所以成就斐然，月球车功不可没。

6. 厄尔·斯威夫特写了一本关于月球车的新书《穿越没有空气的荒野——月球车与登月的最终胜利》。他说："从艾伦·谢泼德的首次飞行任务开始，载人航天计划中的每一次任务，都是为阿波罗的最后三次任务奠定基础。"

7. 斯威夫特说："可以看出美国国家航空航天局集大成，把过去十年在太空中得到的所有智慧都用上了。其惊心动魄之处是其他科学探索无法比拟的。"

8. 特塞尔·缪尔·哈莫尼是华盛顿史密森学会国家航空航天博物馆的阿波罗

藏品馆长。据她说:"尼尔·阿姆斯特朗在月球上迈出的那一小步,实现了阿波罗计划的地缘政治目标,美国国家航空航天局于是力推科学研究。"尽管第一批在月球行走的航天员在着陆点附近采到了样品,但长期以来,科学家们一直希望驾驶月球车沿路勘探能带回稀有的岩石标本。就在阿姆斯特朗和巴兹·奥尔德林首次登月两个月之前,月球车计划终于获批了。

9. 多年来,人们对月球车的模样浮想联翩,但真正在月球上驾车远比听起来要复杂得多。整个20世纪60年代,工程师们研究了对月球车的各种构想:像坦克一样的履带车、飞行汽车,甚至是一个圆滚滚的怪物形状汽车,如斯威夫特描述的那样,"像一个巨大的Tootsie Pop棒棒糖,它的球形舱室带有一条长腿,而这条腿又安装在一个履带轮底上。"最终,工程师们确定选用形似汽车的造型。

10. "当时还有其他一些稀奇古怪的想法,例如选用弹簧单高跷或摩托车——我很庆幸他们最终放弃了这些想法,"缪尔·哈莫尼博士说,"某种程度上,月球车相对更实用。"

11. 月球车也具有典型的美国特色。缪尔·哈莫尼博士说:"月球车裸露的底盘,伞状天线和细辐条式车轮使其外观与市面上的其他车辆判若云泥。而月球车与美国汽车工业的关联以及美国人对月球车的痴狂令人瞩目,这是自'阿波罗11号'以来前所未有的。"

12. 从20世纪60年代的"水星计划"开始,佛罗里达州的一家汽车经销商以1美元的价格将雪佛兰汽车租赁给航天员,这些汽车后来向公众出售。"阿波罗15号"的航天员选择了红、白和蓝色科尔维特跑车。缪尔·哈莫尼博士说:"一张刊登在《生活》杂志上的照片广为流传,照片上那几个航天员与他们经典的美式肌肉车连同月球车合影留念,映衬之下月球车倍儿酷。"她补充说:"那张照片可谓意味深长啊。"

13. 欧文和斯科特与月球车一起抵达月球的消息传回美国,民众立刻就激奋起来。在执行任务的第二天,航天员们驱车前往"斯普尔"环形山,在那里他们发现了一大块白色结晶岩石。这是地质学家梦寐以求的矿石种类,因为从它们身上也许会找到月球起源的线索。

14. 当时,航天员们欣喜万分。斯科特不禁喊道:"哦,天啊!"。欧文也高兴地笑了,忍不住说:"快看它的光泽!"斯科特急忙用无线电向地球报告:"猜猜我们刚刚发现了什么?猜猜我们刚刚发现了什么!我们这次可算是没有白来。"

15. 这块白色的岩石后来被命名为"创世岩",因为最初科学家们认为它是月球形成时期的一块遗留物。

16. 缪尔·哈莫尼博士说:"航天员令人激动的发现,以及他们的宝贝月球车一起,使得诸项阿波罗任务圆满完成,并顺利返回地球。尽管进一步推进探月计划变得越来越复杂多变,这次成功仍为我们提供了一个解决问题的思路。"

17. 斯威夫特指出,当时的一些新闻报道称月球车是"地球上最痴迷汽车的人注定会制造出来的,虽然看起来有点滑稽",不过这款外形怪异的车,其诞生过程绝不是看起来那么简单。

18. 因为月球车与航天员同箭出发,而不是单独运送,其重量必须低于 500 磅(226.8 千克)。同时,月球车要承受相当于其自重两倍的乘员和矿物载重重量。在月球上,月球车穿梭于阳光照射处和阴影处之间,其温差超过 500 华氏度;它还要抵御粗糙的月球尘埃和比子弹速度还快的微流星体的侵扰;它还必须在陡峭崎岖的月球地表上行进,途中遍布山脉、环形山以及松散的砾石和粉末。当年,美国国家航空航天局曾威胁如果通用汽车和波音公司不能完成任务,就会在发射前取消月球车计划,为了赶上最后的阿波罗任务,工程师们争分夺秒,奋力完成了设计。

19. "多亏了通用汽车公司的工程师们,否则月球车根本无缘面世,"斯威夫特在一次采访中表示。

20. 他在书中还解释道,尽管当时预算严重超支、时间大大超期,技术难题成堆,但像米切斯瓦夫·格雷戈里·贝克尔(在波兰长大)和费伦茨·帕夫利奇(出生在匈牙利)这样的移民工程师们仍然坚持了下来。斯威夫特称虽然航天员通常更受关注,但工程师在太空计划中作用重大,贝克尔和帕夫利奇两位先生的例子充分展示了移民这个群体在美国的创新方面做出的重大贡献。

21. 他写道:"在美国的登月竞赛中,无论是美国国家航空航天局内部还是那些负责硬件制造的航天公司里,这些移民工程师——这些碰巧生长于美国之外的美国人,他们的智慧和才干发挥了巨大的作用。"

22. 月球车运抵月球表面,航天员将它展开,但驾驶体验却出奇地别扭。宇航员们把月球车与地球上的其他交通工具进行比较。欧文说这辆车像"一匹暴跳的野马"一样颠簸,斯科特说当他试图以 6 英里(9.6 千米)的惊险时速飞快转弯时,月球车尾部竟然左右摆动起来,就像驾驶一艘快艇一样。

23. 任务管理人员设定月球车最远只能行驶至航天员步行的极限距离,以防发生意外时,航天员必须徒步返回航天器。但随着美国国家航空航天局信心增强,每次执行任务时,"阿波罗号"航天员驾车的距离越来越远。航天员离开月球时,车被留在了登陆点,继续在那里收集灰尘和宇宙射线。绕月飞行的航天器偶尔会拍下它们的照片,在一些照片中,甚至可以看到月球车的行驶轨迹。

24. "航天员发现了更多稀奇的岩块,于是科学家便能就此提出不同类型的问题。"行星科学家芭芭拉·科恩说。她就职于位于马里兰州格林贝尔特的美国国家航空航天局戈达德太空飞行中心,研究这些样本。她还说,有了月球车,航天员可以更专注于科学探索,而不用担心氧气或其他消耗性资源耗尽的问题。"

25. 她回忆起几年前参加美国国家航空航天局的一个模拟任务时,科学家们会穿上航天服,在一个沙漠野外站点进行实验,模拟月球或火星上的情形。她记得参与者正准备采集一块样本时,任务控制人员进行阻拦,要检查他们的生命体征。

26. "我们当时想,'来吧'"她回忆道。"这让我意识到,地质学并不是唯一的主导性因素。有一件事月球车可以为你做到:它可以让你提出不同的科学问题,在特定的地点,更容易找到这些问题的答案。"

27. 创世岩这颗可以追溯到月球最早期的矿石证明了科恩博士的观点。如今科学家们仍然在激烈地争论月球是如何形成的,那时月球上的环境是什么样的,进而也争论地球上的最初十亿年是什么样一种情况。

28. "阿波罗 17 号"执行任务期间送回的样本一直被密封保存,从未与外界接触。科恩博士和其他几位科学家准备打开密封罐。她将研究样品中的惰性气体,以了解太阳辐射如何影响月尘。

29. 凯瑟琳·伯吉斯是华盛顿特区美国海军研究实验室的地质学家。她将研究这些原始样本,以测量太阳风辐射如何影响月尘内氢氦含量。航天器可以从轨道上探测到月球上的氦,但是科学家们仍然不知道氦含量是如何随着月球地形变化的。她说:"如果没有这些样本来证实,这个问题会一直悬而未决。"

30. 未来的任务可能会使用月球上的氦,特别是一种叫作氦-3 的变体,作为核反应堆的燃料来源。这意味着半个世纪前被第一代月球车发现的材料,将会成为新一代月球车的动力来源。

31. 在研究这些原始样本时,不少科学家希望能有一批新的样本。这些新样本由新一代航天员送回地球,或由新一代月球车收集。今年五月,通用汽车公司宣布与洛克希德·马丁公司合作,为美国国家航空航天局阿尔忒弥斯计划制造一款新型月球车。按照该计划,美国航天员将在十年内重返月球。

32. 虽然由不同团队打造,而且中间相隔了几十年,但第一代火星探测车无疑受到了月球车的启发,尤其是"旅居者号"这辆第一个登上其他行星的探测车。美国国家航空航天局的火星探测车是在其喷气推进实验室制造出来的,那里的工程师们设计了柔性框架的六轮火星车,与通用汽车的早期设计有相似之处。斯威夫特先生表示:"我坚信火星车的设计与通用汽车早期设计灵感一脉相承。"

33. 如今,科学比地缘政治更能推动美国国家航空航天局的发展,但航天局坚持推动开展人类太空旅行的原因远远不止是勘探岩石。缪尔·哈莫尼博士说:"阿波罗的月球车及其更新换代的新车型代表了一种探险精神。"

34. 哈莫尼说:"的确,阿波罗计划的重要成果体现在科学发现上,但了解公众的关注焦点也很重要。月球车引发的公众热情与人类航天的吸引力性质一样,即可以有机会亲眼见证公众的喜悦激动以及身临其境般的参与感。"

35. 另外,驾车穿越月球的冒险之旅,是有史以来最伟大的驾车之旅,令人难以抗拒。

36. 无论是过去还是现在,"公众的注意力都不在来自月球的样本和材料上,而在月球车上。"哈莫尼说。

【注释】

① 本句话中有两个否定成分,即"be not about to"和"without",主体都是"Dave Scott"。如果将两个否定都翻译出会略显赘余,可仅保留一处否定,将后一个否定的

关注主体由人转向物,更加通顺。

② 此处补出"首开先河",使用了汉语四字格,增强可读性。

③ "an unscheduled pit stop"为名词性短语,翻译时根据中文的表达习惯可译为副词+动词结构。"unscheduled"译为"出乎意料"。

④ "crater"在地球上指火山口,在月球上统称为"撞击坑",即碗状凹坑结构的坑,也叫"月坑"。撞击坑主要由流星体、陨星撞击或者月球火山爆发形成。本段出现的"lava rock"是火山熔岩,"Seatbelt Rock"是一种玄武岩,根据岩石种类可以判断是火山口。

⑤ 此处的"train"不可以翻译成训练,"train"有在大学学习过某方面知识的含义。

⑥ 长句需要适当增词捋顺句子的逻辑关系。"when"在此句中是"在……之后"的意思,使用"等到……后……才……也,而……"可以让句意更清晰。

⑦ "NASA"是美国国家航空航天局的缩写,在文章中第一次出现时按照全称译出,在后面用括号形式标出缩写,再次出现可直接使用缩写。

⑧ 多个修饰词并列时,不要把这些词全放在名词前,可以根据其性质和特点分出层次。

⑨ "take all of that collected wisdom"如果直译较为生硬,将其译为"集大成",既包含了原文"其中各个方面,达到相当完备的程度"之意,又有汉语简洁明了的特征。

⑩ 此处可采用反说正译的技巧,译文会更加清晰。

⑪ 此句中"excitement"的主体是民众,为了使译文更加简洁,可以将后半句中的主体直接换成民众,减少重复。

⑫ "on …'s wish lists"直译过于生硬,可以根据句意将其转换为其他形式。

⑬ 此句如果直译人称会出现两次,重复赘余,可保留其含义,巧妙简化。

⑭ "Genesis Rock":创世岩,"阿波罗 15 号"航天员带回的月球岩石样本。

⑮ "excitement"是抽象名词,根据文章大意将其具体化为"令人激动的发现"。

⑯ 此处根据上下文需要增译出阿波罗项目圆满完成,顺利回到地球这一情况。

⑰ 此处的翻译需要结合下文,月球车的使用条件苛刻,制造难度大,所以月球车的诞生艰难,绝不是简简单单就能造就的。

⑱ "horseless carriage":汽车早期的名称。

⑲ 此处如果直译成"科恩博士是……的几位科学家之一"较为生硬,而且前面定语过长,翻译时可以把包含关系调整成并列关系。

⑳ "its modern successors"指的是代代更新的月球车,需将实际含义译出。

㉑ "be engaged with"指的是公众对于这一事情的关注。

㉒ 为避免重复,主句中两个"appeal"尽量不要都译成"吸引力",可以根据在文章中"appeal"的具体含义译出。从句中的"their"指的是前面提到的公众,不可以模糊

翻译成"他们的"。

㉓ 文章中几处的翻译都使用了汉语四字格：

the first people	首开先河
unscheduled	出乎意料
valuable	价值非凡
crowning achievement	成就斐然，功不可没
had been imagined	浮想联翩
it looked like no car	判若云泥
on geologists' wish lists	梦寐以求

英文中表达如果按照其形式直译成汉语会略显啰嗦，巧妙使用四字格可以简洁明了地表达原文含义。同时，一些形容词使用四字格翻译会更具有可读性。

课后练习参考答案

第 1 章

第 2 节

一、

1. 水上飞机

2. 航空电子（aviation electronics）

3. 涡扇（turbine fan）

4. 教练机

5. 全角视野

6. 空中交通管制（Air Traffic Control）

7. 无内胎轮胎

8. 齿轮箱,变速箱

9. 起落架

10. 旁通活门

11. 航空电子专家

12. 油门杆

13. 外场支援

二、

1. 这要求对油箱系统的外观设计进行识别,进而推进油箱点火源的发展。（identification 和 development 名词化结构）

2. 多年前就有能够以一定精度来记录最大压力和出膛速度的仪器。（capable of …形容词短语作后置修饰语）

3. 紧密性的部分好处还在于,可以通过半封闭循环配置实现的更高的总循环压力比,大大降低低压组件和休整器的重量。（长句；achievable with …形容词短语作后置修饰语；leading to…非限制性动词）

4. 多年来,美国国家航空航天局（NASA）和科学界一直争取实施火星样本返回（MSR）任务。已经进行了大量研究来评估该任务的任务架构、技术需求、发展计划和顶层要求。（缩略语 NASA、MSR；现在完成时,表示到目前为止发生的行为、对现在产生影响的行为）

5. 基于伽利略的研究成果,人们逐渐认识到,存在着支配物体运动的普遍规律,这些规律不但适用于地球上的运动,也适用于宇宙空间的运动。（同位语从句作修饰语,和先行词分割开；governing the motion of bodies 分词结构,作后置修饰语）

6. 电力推进结构很有吸引力,因为它为中型渐进型一次性运载火箭(EELV)提供了一个单一的发射解决方案,其超长的发射窗口期长达几个月。(缩略语 EELV;独立结构 with very large launch windows)

7. 如果航空航天工业需要转向非常规布局,在设计周期的早期阶段对飞机飞行行为进行准确评估是至关重要的,这不仅是为了防止造成很大损失的最后一分钟返工,也是为了评估它们的可行性。(if 引导的条件句;to accurately evaluate……不定式短语作后置定语)

8. 2010 年,美国国家航空航天局(NASA)设立了美国国家航空航天局人类健康和行为中心(NHHPC),为交流创新方法的最佳做法提供了一个论坛,促进跨部门发展伙伴关系,以解决太空和地球上的医疗、环境和人类行为问题。(缩略语 NASA、NHHPC;长句)

9. 1887 年进行的米契尔森-莫利实验旨在确定地球在以太中的运动。以太是一种想象中的物质,它绝对静止、无形、无色、无味。但它在宇宙中却无处不在,这样光才能通过介质传播。(长句;被动语态;同位语:a presumed absolutely motionless, invisible, tasteless, and odorless substance 为 the ether 的同位语)

10. 鉴于空间飞行人员的多种类型和不断增加的飞行时间,需要更先进的医疗保健和环境卫生系统,并采用新的方法和技术来改善飞行中的医疗问题。(given 引导的条件状语从句;被动语态;长句)

第 3 节

1. 附近的天体一旦落入黑洞,就销声匿迹,永无踪影。

2. 因此,不管你是进行基本设计,还是自己创造新式样,能够告诉你风筝是否称心如意的唯一方法就是:放飞。

3. 他观察到,木星的卫星似乎并非匀速地绕过木星背面,并不像人们预期的那样,以恒定的速度环绕木星运行。

4. 总之,他(爱因斯坦)用新的科学架构替代了 19 世纪科学界引以为傲的结构。后者在世纪之交曾遭到众多轰动一时的发现的重创。而他的新的科学架构的基础和框架却非常坚固,足以抵抗将来可能来临的最剧烈的风暴袭击。

5. 从人造卫星上把一个装有无线电发射机的橡皮球抛下,向行星(火星)表面坠落。无线电发出橡皮球坠落时减速的信号,科学家们就能由此计算出大气的密度。甚至还可以把一个装载着科学仪器的密封舱(从卫星上)抛落到火星表面上去。

第 2 章

第 1 节

1. 这是一架 20 倍的双镜头显微镜。

2. 橡胶不硬,受压就会变形。

3. 弧形曲面反射器具是太阳能装置的关键元件。

4. 合金是一种介于混合物和化合物之间的中间物质。

5. 软件包是为了满足某一用户的特定需求而开发的。

6. 有些反应发生得很快,比如酸碱溶液之间发生的反应。

7. 位于德比郡的铁路技术中心的英国铁路工程师大大减轻了列车的重量和牵引阻力。

8. 私人飞机主要用于休闲飞行,通常型号为单引擎单翼机,搭载固定式起落架。

9. 大型导弹通常无导轨发射系统,不论是系装还是竖立在大发射场上,都受风的影响。

10. 过滤器拆卸、检查及清洗的次数主要取决于飞机的运行状况。

11. 其中两种分别叫作丁烷和丙烷的气体,有时可以装入钢瓶,在没有管道燃气的地方使用。

12. FLightstats 全球飞行跟踪网站可以帮助你查看你的航班是否延误或取消,并在地图上跟踪实时位置。

13. 你可以松缓地半躺在扶手椅内,享受几小时的飞行时光。真正会享受的人还可以在某些航班上看看免费电影,呷呷香槟。

14.

原　　文	译　　文
Solidangle	立体角
Solid body	固体
Solid line	实线
Solid color	单色
Solid injection	无气喷射

第 2 节

1. 公众广泛参与,委员会也很好地利用了公众的评论。

2. 他们特别注意提高磁控管和多腔速调管的质量。

3. 毫无疑问,火星是内太阳系最具科学意义的目的地。

4. 天文家宣布,可能在太阳系之外发现了第一颗已知卫星。

5. 通过建立目标,关于目的地、探索策略和运输架构的问题才可以按照逻辑顺序进行。

6. 其新颖设计中融合的技术和创新,使战斗机的性能有了显著提高。

7. 所提供的支持包括各种风洞和旋转隧道测试、跌落模型测试和模拟器测试,以支持飞行控制系统的开发。

8. 然而,国会指出这些飞机中只有一架将作为空军所说的两用战斗机(DRF)投入生产。

9. 在这张哈勃望远镜拍摄的照片中,来自遥远星系的光线在重力的作用下绕着星系团的中心弯曲,形成了所谓的爱因斯坦环。

10. 通过引入热交换器可以显著提高循环热效率,但发动机的总重量也随之增加。

11. 通用动力公司想要说服空军 F-16XL 战机与标准 F-16 战机的生产型号互为补充,且无须与机型庞大的 F-15 战机竞争。

12. 在 YF-16 轻型战斗机技术演示计划和后续 F-16 战机全面发展计划期间,通用动力公司和美国国家航空航天局建立了高效的合作关系。

13. 对于美国来说,高速地面运输是一种新的运输方式。路程在 100～600 英里(160～960 千米)之内,这种运输方式比飞机、汽车更有潜在的优势。

14. 在 20 世纪中期以前,实用的"计算机"不是机器,而是借助简单工具(如算盘或计算尺)进行数学计算的人。

第 3 节

1. 温度大大高于设定值。

2. 橡胶性软,受压即变形。

3. 瞳孔可随光线强弱而扩大或缩小。

4. 即使使用最高放大倍数的显微镜,原子还是小得看不见。

5. 重约 1 吨的电磁铁能吸住 7～8 吨的钢铁。

6. 大多数彗星都非常黯淡,远非裸眼可见。

7. 多年来,煤炭和石油一直被视为交通运输的主要能源。

8. 人类已开启太空时代,昔日的幻想似乎就要变成明天的现实了。

9. 在一场激烈的大规模遭遇战中,现代护卫舰或驱逐舰可以在头几个小时将舰载地空导弹消耗殆尽。

10. 这次试验的结论是,只要首次拦截及时,就有 99％的把握使导弹丧失进攻能力。

第 4 节

词的增译

1. 销子安装后,将铰链两端紧固,以免销子脱落。

2. 晶体管可以使原本体积大的设备变得更小。

3. 假如这一反应需要数小时而不是几秒钟,那么燃料成本就太高了。

4. 我们的太阳只不过是一个普普通通、平均大小的黄色恒星,邻近其中一个螺旋臂的内边缘。

5. 由于缺乏足够的知识,在标本分类时存在将蝴蝶和飞蛾混为一谈的情况。

6. 技术的兴起,特别是信息和通信技术的发展,正在彻底改变人们的生活方式。

7. 电的生产和储存的基本方法有三种:用发电机、用电池、用蓄电池。

8. 当把一根铜线的两端连接到一种叫作电池的装置上时,就会有稳定的电流流过该铜线。

9. 这项发明的目的是提高内燃机的动力、增强适应性、降低油耗和减少成本。

10. 对于这个提法的定义绝非唯一，而是各项因素的组合，如全新的或更新换代的铁轨、车辆、实际运营，这些都促成了高速铁路的运营。

11. 这也许有悖于我们的直觉，但指示牌确实会让我们慢下来。你不禁止步细看。所以亚特兰大机场尽量减少使用指示牌。但是这不只是降低我们的实际移动速度，还减缓我们的理解认知。

词的省译

1. 这样的发动机称为内燃机。

2. 内外圈是根据尺寸分类存放的。

3. 温度可能达到 700 摄氏度，表面氧化变得至关重要。

4. 要彻底清除消耗臭氧的化学品还需要时间。

5. 如果空气通过裂缝进入灭弧室，真空断路器就会失灵。

6. 各种材料可按物理性质加以区分。

7. 重元素的原子核分裂会释放大量能量。

8. 从那时起，用电时电表不是逆转就是完全停转。

9. 物质的熔点就是从固态变为液态时的温度值。

10. 要想进行基本设计，必须了解整个设计顺序。

11. 电流通过电阻时，电阻的两端就会产生电势差。

12. 每一个臂状结构都由塑料薄膜制成，机器人或航天员将其铺展在月球表面。

13. 那个时候，人们像着了魔似的想要了解地球，想要确定地球的年龄和体积，悬在宇宙的哪个部分，是怎样形成的。

第 5 节

1.

space tracking and data network　空间跟踪和数据网络

Polar-orbiting meteorological satellite　极轨气象卫星

Quick-look photographic reconnaissance satellite　快查型照相侦察卫星

radio transmission photographic reconnaissance satellite　无线电传输型照相侦察卫星

2.

Prandtl number 普朗系数

parametric Wiener filter 参数维纳滤波器

Poisson's coefficient 泊松系数

Mercator's plotting chart 麦卡托航线测绘图

Monte Carlo restoration 蒙特卡洛复原

Apollo-Soyuz test project (ASTP) "阿波罗-联盟"试验飞行计划

3.

OMS engine 轨道机动系统发动机

portable LSS 便携式生命保障系统

NiFe accumulator 镍铁蓄电池

NAS screw 国家航空标准螺旋桨

DAP rate dead band

数字式自动驾驶仪角速度死角

homomorphic DPCM　同态差分脉码调制

4.

taxi post 滑行起飞跑道停机位置

propeller clearance 螺旋桨与机身的安全间隙

propulsive lift 升力发动机的垂直推力

throat of the nozzle 喷管喉部临界截面

pro rata charter 按比例分成的包机航班

profile thrust mode 按翼型自动变化推力模式

第 3 章

第 1 节

1. 哈勃望远镜的探测速度是天文学历史上任何望远镜都无法比拟的。

2. 我们能否在合理保证人身安全的情况下进行探索？

3. 然而,经过了几个世纪,直到现代,我们才认识到星系在宇宙中的作用和重要性,其中包括人类生存的银河系。

4. 通过评估那些负责研究的领导者在特定领域可能产生的影响,来评定创新价值。

5. 最初,第二阶段的规划重点是建设并使用多功能机载试验台进行研发。

6. 海面船舶检查图像和回收物品后,并未发现任何与马航 370 相关的物品。

7. 哈伯通过开发利用空气中的氮气而彻底扭转了局面。

第 2 节

1. 在所有情形下,破断面的两侧都发现有放射性。

2. 预计过了 2020 年哈勃望远镜仍运行良好。

3. 他建议必须彻底解决飞行员失能的问题。

4. 图中方框是银河系的一个小区域,在其左下角突出显示了六个星团。

5. 他特意指出,如果第二名飞行员被地面站的人员替换,会对安全风险造成重大影响。

6. 虽然该战机最初设计用途为轻型空战战斗机,但后来却越来越多地承担了地面攻击任务。

7. 土星的第二大卫星土卫五(直径 949 英里,或 1528 千米)被图像的右边缘一分为二。

8. 这种机型的外翼及水平稳定翼（与较小的喷气机机翼大小相当）将由碳纤维复合材料制成。

9. 流星光谱一般由地面仪器或飞行仪器记录，然后与合成光谱做比较，以确定元素丰度和温度。

10. 还测量了太阳光谱辐照度（SSI），以确定这种能量在不同波长之间的分布以及大气中吸收该能量的具体位置。

11. 天气预报员越来越频繁地使用夜光数据，因为可见光红外成像辐射仪（VIIRS）可以用来观测夜间被月光和闪电照亮的云层。

12. 飞行测试计划以惊人的速度进行，并于 1975 年 1 月 31 日完成。两架实验样机完成了 347 架次测试，飞行时长共计 439 小时。

13. 全国各地的机构都使用其洋面风速和风向的测量数据来进行天气和海洋预报，监测飓风和热带气旋。

14. 这张经过着色处理、大幅增强的图像显示了土卫二羽流中规模较大的阴暗部分，它区域辽阔，是水流从南极裂缝中喷出时产生的。

15. 据设想，光纤传感系统（FOSS）可以收集空气动力对飞机影响的实时信息，并将这些信息直接循环导入飞机的控制系统中，以提高燃油效率、安全性和乘客的乘坐舒适度。

16. 然而，（我们）应该注意到在关于飞行员失能的对话中，有几条评论表明一些与会者认为该问题被夸大其词了。

第 3 节

1. 目前尚不清楚单人操作是否可行。

2. 纵观诸多方面，从冰山模型的角度来考虑这一问题很有启发性。

3. 这也是德莱顿飞行研究中心最初成立的原因。

4. 他表示忽视直接枢纽区以外的活动是目光短浅的。

5. 可以确切地说，F-16XL 战机原型配置的诸多要素直接源于与美国国家航空航天局联合研究的成果。

6. 需要注意的是，高度自动化的功能配置在某些指标上表现非常好，而在另一些指标上则表现欠佳。

7. 我们认为这种电子监控装置不太可能被水下探测器检测到，因为所使用的超高无线电频率，一旦进入水中会迅速衰减。

8. 例如，我们有理由相信，在这些情况下（而非飞行员失能的情况下），自动化系统作为飞行员的备用设备可能会获得更广泛的认可。

9. 飞行员向空中交通管制中心通报飞机是由多名机组人员还是单个飞行员操作，这可能会成为一项重要要求。

10. 正是在第二阶段，技术概念达到了就绪、有效和可靠的状态，这样，私营企业和融资方就可以承担随之而来的风险。

11. 约翰逊航天中心和德莱顿的工程师们一致认为,如果有合适的测试飞机,探索航天飞机轮胎和车轮在现实条件下的实际极限和失效模式将大有裨益。

12. 由于在航天飞机所处的极端温度和环境下保护轮胎和齿轮非常困难,轨道飞行器只配备了四个小轮子,每个主起落架上两个。

13. 但是随着美国的研究重点转向航天领域,以及运输机在配置和速度上都趋于稳定,再加上军方新型尖端战斗机的产量减少,像德莱顿这样的地方的价值,外界更难看到。

14. 很明显/显而易见,虽然美国可能是第一个实现飞机动力驱动、可控飞行的国家,但在德国、法国和英国,得益于政府资助的研究成果,欧洲飞机制造商的实力远远超过了美国。

15. 但正是通过埋头苦干,有关各方才明白了跨声速和超声速情况下的致命危险,也了解到了设计缺陷及其补救措施,而这挽救了无数追随第一批勇敢的探险家脚步的飞行员的生命。

第 4 节

1. 检测大气中的臭氧及其化学成分对了解臭氧层的耗损至关重要。

2. 在评估国际空间站研究成果的经济价值时,采用这种类比方法可以避免一些陷阱。

3. 国际空间站内有 30 多个具有不同光学特性的窗口,为研究人员提供了许多观察机会。

4. 尽管国际空间站服务环境研究和可视化系统旨在改善自动图像捕获和数据传输,但是实验中拍摄的图像也为环境科学家、灾难救援人员和其他地球上的用户提供了便利。

5. 轨道传感器系统获得的遥感数据已成为确定自然灾害损害程度的重要工具。

6. 在华盛顿特区面对公众举行的"大辩论"中,天文学家哈洛·沙普利和希伯·柯蒂斯就旋涡星云是否从属于银河系提出了各自的主张。

7. 大气-空间相互作用监测器(ASIM)测量的结果可以用来完善当前的大气模型,包括与气候学相关的预测,并且可以加深对这些事件的物理性质以及其与闪电关系的了解。

8. 在海面搜寻过程中,飞机报告了漂移搜寻区域的一些漂浮物体,包括木制托板和捕鱼设备。

9. 天文学家利用哈勃望远镜敏锐的观测力来探测可见宇宙的极限,从而发现了在大爆炸中宇宙诞生后不久就存在但从未观测到的物体。

10. 从 2001 年开始,在航天飞机前往国际空间站的几次飞行任务中,安进公司利用微重力环境进行了三种药物的试验。

11. 1923 年,埃德温·哈勃使用位于加州威尔逊山顶上的世界最大望远镜测定了(地球)与仙女座大星云的距离,结果发现,它与我们相距甚远,并不属于我们的星系。

12. 臭氧层就如同地球的宇航服,通过吸收太阳最有害的紫外线来保护生态系统和人类。

13. 成立航天材料部门的目的是整合科技进步成果来推动人类航天探索的发展,应用行星研究成果,以及寻找缓解方法以实现成功的太空飞行。

14. 火山灰由微小的玻璃和岩石碎片组成,会磨损发动机涡轮叶片,并可能熔化在叶片和发动机其他部件上,从而损坏(发动机),甚至导致发动机熄火,进而危及飞机的完整性和乘客的安全。

第5节

1. 1915年,阿尔伯特·爱因斯坦提出了一个革命性的设想,即大质量物体会扭曲空间。

2. 繁星点点的星空似乎是一块宁静的幕布,布满了明亮、不变的灯塔。

3. 这片星云为我们的星系提供了更多物质,为新恒星的形成提供了条件。

4. 和所有的望远镜一样,哈勃望远镜就像一台时间机器,天文学家可以用它来观察过去。

5. 变压器不能使用直流电,因为会烧坏变压器内的电线。

6. 像煤和木材这样的固体燃料只能在表面燃烧,因为表面能接触到空气。

7. 最初的临时飞行研究站最终成为永久的飞行研究中心。

8. 另一种解释是,中子星可能会喷出一股高能气流,猛烈地撞击星际空间中的气体。

9. 因其沿着麦哲伦星云运动的方向向前延伸,天文学家称它为"引导臂"。

10. 尽管该星系距离我们250万光年,但哈勃望远镜的观测结果显示该星系盘中有超过1亿颗独立的恒星。

11. 分析表明,飞机在这一段飞行中的燃油消耗率高出预期,因此缩短了其最大航程。

12. 最初,管理层支持远程工作的理念,但对中心的大部分工作可以远程完成的想法表示怀疑。

13. 哈勃望远镜使用仪器阻挡星光,使恒星附近较暗的特征和物体明晰可见,从而增进了我们对恒星周围环境的了解。

14. 委员会因此确定了五个问题,这些问题可以构成美国载人航天计划的基础。

15. 引力透镜的强度如此之强,以至于伊卡洛斯(Icarus)在恢复到最初的光泽之前,曾短暂地飙升至其真实亮度的2000倍。

16. 每个"眼睛"都是一个星系明亮的核心,而所有蓝色区域都是年轻恒星因碰撞而迸发出生命的地方。

17. 这个扭曲的星系被命名为NGC 3256,是大约5亿年前两个螺旋星系碰撞的产物。

18. 尽管天文学家已经找到了稍远一些的其他明亮星系,但这个天体是一类更小、更暗淡的新生星系,此类星系在早期宇宙中肯定很常见,但基本上没被人类探测到过。

19. 声呐浮标配备一个水下探测器,该探测器可下探到浮标下方300米深处;浮标还配备一个无线电发射器,当声波数据在有效范围内时,可以将声音数据传送回飞机进行处理。

20. 但是,如果美国想要继续探索外太空领域,那么用于进入其他行星大气层和行星表面的新型创新飞行器和系统,几乎肯定会在美国国家航空航天局的未来研究中占据重要地位。

第6节

1. 解决这个问题最简易的方法是发明一种发动机,曲轴/传动轴可以位于其顶部,这与汽车发动机的曲轴总是在其底部的设计完全不同。

2. 这项研究得到了"月宫365"的支持。"月宫365"是一个历时370天的多机组人员封闭实验,该实验在一个名为"月宫一号"的基于地面实验生物再生生命保障系统(BLSS)的平台上进行。

3. 这一点出乎意料,因为如此大量的硫只能源自前几代恒星,而这些恒星不曾存在于我们银河系之外的原始氢云中,也不曾存在于未能形成、无恒星的星系的残余中。

4. 哈勃望远镜和其他望远镜的观测结果表明,像银河系和仙女座这样的螺旋星系是在矮星系合并的过程中形成的;并且直到现在,它们还在吸引并吞并这些较小的星系卫星而不断变大。

5. 从商业航空运输初期开始,常规的管翼式设计就一直是标准的飞行器结构布局,而如今研究人员要进一步确定,除了管翼式设计,其他各种先进的飞行器机构布局有哪些潜在优势。

6. 美国国家航空航天局的研究人员认为,在2026年至2050年期间,由于美国国家航空航天局及其行业伙伴在环境责任航空(ERA)项目中开创了"绿色"航空技术,各商业航空公司可以节省高达2500亿美元的成本。

7. 尽管所有"配置"研究飞行器都表现出独特而怪异的特点和问题,但它们都对航空航天工业做出了重要贡献:例如,在接近0.9马赫时,X-4的三个轴都产生剧烈振荡,这让研究人员相信,半无尾结构完全不适合跨声速或超声速飞机。

8. 如果由于缺乏资金或得不到支持,导致德莱顿的优秀人才离职,或者中心的测试机被送去销毁,那么德莱顿的各项功能都将受到影响,包括组织成功的模拟飞行研究,支持业界推进技术进步的各项工作,帮助美国国家航空航天局开发新的航天器设计,甚至是支持开展使现有飞机或航天器更安全的研究工作。

9. 通用动力公司希望通过独立研发一种实验样机来获得空军对研发和生产新

型飞机的支持。这个样机能够以较低成本验证跨声速/超声速巡航和机动性概念,还将提高空对地作战能力,不过这个新机型与基本型 F-16 战机有很多相似之处。

10. 该中心规模小,注重实效,管理和运营方式超脱、灵活且创新,这都有助于培养一个既有技术热情又有技术灵活性的工作团队,这些特点使该中心在承接各种各样的研究项目和重点工作的同时,能够不断适应过去 60 年来时代和研究重点的变迁。

第 4 章
第 1 节

1. 美国国家航空航天局新投入使用的太空望远镜首次拍下星光,甚至还为它巨大的金色镜子拍了一张自拍照。(拟人)

2. 就像摆渡者可以让船垂直于水流方向行驶一样,行星也能在只有恒定侧向气流的情况下自由进出。(类比)

3. 他(开普勒)假设太阳绕轴自转,形成一个动力漩涡,推动行星绕其转动。(暗喻)

4. 美国国家航空航天局刊出了这张"自拍照",以及每部分镜子反射出的星光的拼接图。这十八点星光像明亮的萤火虫,在漆黑的夜空中飞舞。(明喻)

5. 科莫萨博士说:"恒星经得住轻度拉伸,但这颗恒星被拉伸到了强度极限。这颗可怜的小星星只是出现在了不该出现的地方。"(拟人)

6. (内燃机)恰当的点火时机极其重要,极高的精度也是如此,对喷油泵与气阀某些部件的研磨精度应小于两万分之一英寸。(消极修辞)

7. 配备自动安全保障装置的飞机,其结构部件的疲劳断裂,只要不是过于频繁地发生危及飞机安全、缩短飞机使用年限,或因维修过多降低飞机利用效率和经济效益,是完全可以接受的。(或:配备自动安全保障装置的飞机,其结构部件发生疲劳断裂是完全可以接受的,前提是不要过于频繁地发生危及飞机安全、缩短飞机使用年限,或因维修过多降低飞机利用效率和经济效益。)(消极修辞)

8. 在风扇的空气入口处,通常装有活动的百叶窗,它受恒温器控制能自动地开闭,尽可能地保持气缸温度稳定。发动机大负荷运转时,吸入的空气就多些,空转时,吸入的空气就少些。(消极修辞)

9. 与极限荷载有关的是保险系数和终极系数,保险系数是选定用来保证即使极限荷载加在飞机上,飞行也不会受到损害;终极系数则是用来预防结构强度可能发生的变化。(消极修辞)

10. 紧接着又发射了其他一些越来越精密的通信卫星,尤其是"晨鸟一号"——这是第一颗商业通信卫星,于 1965 年发射,现在盘桓于大西洋上空大约 22300 英里(35 680 千米)的地方,定期在欧美之间传送电话和电视信号。(时态)(消极修辞)

第 2 节

1. 假如月球上真有居民的话,他们就会看到地球像悬挂在天空中的一面巨大的镜子,反射着太阳光。

2.他们曾研究了一些奇怪的方程,并思考这些方程究竟代表了什么,而如今我们已经知道这些方程描述的是一种叫作"黑洞"的、无法逃离的一小块空间。

3.我们可以通过研究弹丸重心的移动来研究弹丸的运动,因为弹丸的质量通常被认为集中在其重心上。

4.同是臭氧,在平流层它能阻挡紫外线,保护人类;但在地面,吸入臭氧会严重损害呼吸系统。

5.近距离看,这个星系中似乎遍布着年轻、明亮的蓝色恒星,但由于它们发出的光在宇宙膨胀的过程中被拉长(变得偏红)了,整个星系看起来是红色的。

6.根据该原则,隐形飞机的设计必须实现其在所有针对它的监测系统上的暴露距离大体相同。如果某飞机在距离雷达 5 英里(8 千米)时不会被发现,但在距离光学探测器 10 英里(16 千米)时被捕捉,那么建造这样的飞机就失去了意义。

7.这种量子存储器的纠缠光源选择较为灵活——确定性纠缠光源也是选项之一,同时还能保持多模式复用能力,因而在量子中继应用中更为高效。

8.尽管罗杰斯舰长知道这架飞机显然用加密的军用频率发射过某种信号,他却不知道这架飞机曾与民用指挥塔有过常规飞行指令对话,并且是用公开的无线电频道播送的。

9. 2017 年,国际空间站安装了美国国家航空航天局研发的中子星内部组成探测器(NICER)。2019 年,这台 X 射线望远镜探测到一颗质量为太阳 1.4 倍、距离地球 1000 光年、直径约为 26 千米的中子星 J0030。

10.最近,两个独立团队利用 NICER 的数据,对距离地球 3000 光年的另一颗中子星 J0740 展开了相同的分析研究。两个团队算出的直径基本一致,分别为 24.8 千米和 27.4 千米,只有几千米的误差。这一结果令人惊讶,要知道,J0740 的质量是太阳的 2.1 倍,是已知质量最大的中子星,比 J0030 还重 50% 左右,但两者的直径却几乎一致。

参考文献

中文文献

[1] 曹明伦.形具神生,神形兼备——谈英语长句的汉译[J].中国翻译,2015,36(06):103-105.

[2] 曹世超,李照宇,苏伟,等.铁道工程科技论文翻译浅谈[J].中国科技翻译,2011,24(02):13-16.

[3] 陈桂琴.科技英语长句翻译方法例析[J].中国科技翻译,2005(03):5-7+65.

[4] 陈嵘.篇章语言学理论范式的演进:从"规约论"到"自然论"[J].语言研究,2021(09):26-34.

[5] 陈新.英汉文体翻译教程[M].北京:北京大学出版社,1999.

[6] 程洪珍.英汉语差异与英语长句的汉译[J].中国科技翻译,2003(04):21-22.

[7] 党争胜.结构分析翻译法初探——浅论英语长句的汉译[J].外语教学.2006(7):64-66.

[8] 丁树德.科技术语翻译中的概念转换[J].中国翻译,1996(05):25-26.

[9] 樊才云,钟含春.科技术语翻译例析[J].中国翻译,2003(01):59-61.

[10] 范武邱,杨寿康.科技翻译的虚实互化[J].中国科技翻译,2001(02):1-4+54.

[11] 范瑜,李国国.科技英语文体的演变[J].中国翻译,2004(9):86-87.

[12] 方梦之.科技翻译:科学与艺术同存[J].上海科技翻译1999(04):32-36.

[13] 方梦之,范武邱.科技翻译教程[M].上海:上海外语教育出版社,2008.

[14] 冯庆华.实用翻译教程[M].上海:上海外语教育出本社,1997.

[15] 冯世梅,杜耀文.现代科技英语词汇与翻译[J].中国科技翻译,2002(04):52-54.

[16] 傅敬民,喻旭东.大变局时代中国特色应用翻译研究:现状与趋势[J].上海大学学报(社会科学版),2021(7):128-140.

[17] 付娆.非谓语动词的翻译策略探究[J].国际公关,2020(02):115.

[18] 傅勇林,唐跃勤.科技翻译[M].北京:北京外语教学与研究出版社,2012.

[19] 高博.变译理论视角下的技术文本翻译探究[J].宁波广播电视大学学报,2019(12):23-26.

[20] 耿智,马慧芳.认知-功能视角下英语定语从句的翻译[J].上海翻译,2015(01):42-45.

[21] 龚益.社科术语工作的原则与方法[M].北京:商务印书馆,2009.

[22] 古今明.英汉翻译基础[M].上海:上海外语教育出版社,1997.

[23] 郭富强.意合形合的汉英对比研究[D].华东师范大学,2006.

[24] 郭建中.科普与科幻翻译[M].北京:中国对外翻译出版公司,2004.

[25] 郭蔓娜,肖红英.浅析非谓语动词作修饰语时的英汉翻译[J].太原教育学院学报,2006(03):72-74.

[26] 郭树林,郭剑.功能对等理论视角下的科技术语翻译[J].中国科技术语,2015,17(03):28-31.

[27] 韩其顺,王学铭.英汉科技翻译教程[M].上海:上海外语教育出版社,1990.

[28] 何其莘,仲伟合.科技翻译[M].北京:外语教学与研究出版社,2012.

[29] 洪忠民.科技术语翻译的规范与统一[J].徐州工程学院学报,2006(04):54-56.

[30] 胡显耀,曾佳.翻译小说"被"字句的频率、结构及语义韵研究[J].外国语(上海外国语大学学报),2010,33(03):73-79.

[31] 胡壮麟.系统功能语法概论[M].长沙:湖南教育出版社,1996.

[32] 黄浩森.消极修辞与积极修辞——学习《修辞学发凡》札记[J].当代修辞学,1982(01):25-26.

[33] 黄华远.科技翻译的表达问题[J].中国科技翻译,1995(02):15-19.

[34] 黄俐丽.科技英语的翻译技巧[J].浙江工商职业技术学院学报,2003(03):88-90.

[35] 黄湘.科技英语汉译的词义引申[J].中国科技翻译,2001(02):29-31.

[36] 霍金.时间简史[M].许明贤,吴忠超,译.长沙:湖南科技出版社,2002.

[37] 矫天顺.科技英语中一词多义及其翻译技巧[J].中国翻译,1989(06):31.

[38] 金微.知其然,亦应知其所以然——论英语关系分句(定语从句)及其汉译[J].中国翻译,1991(05):14-16.

[39] 康志洪.科技翻译[M].北京:外语教学与研究出版社,2021.

[40] 匡丽.航空航天科技英语术语翻译研究[J].时代农机,2018,45(12):77-78.

[41] 冷冰冰.MTI科技翻译教材应贯穿的五元次能力[J].上海翻译,2016(1):47-54.

[42] 李丙午,燕静敏.科技英语的名词化结构及其翻译[J].中国科技译.2002(2):5-7.

[43] 李怀先.科技英语汉译琐谈[J].中国科技翻译,1988(01):27-32.

[44] 李鲁.试论科技翻译的语义引申[J].中国科技翻译,1995(03):6-8+29.

[45] 李清.科技英语翻译中语言/技术逻辑判断与教学[J].教育与现代化.2009(4):38-42.

[46] 李庆新.英语语法理论与实践[M].北京:光明日报出版社,2013.

[47] 李汝辉,吴一黄.活塞式航空动力装置[M].北京:北京航空航天大学出版社,2008.

[48] 李瑞晨.英汉俄航空航天词典[M].北京:北京航空航天大学出版社,2002.

[49] 李雪,常梅.科技英语翻译教程[M].北京:经济科学出版社.2020.

[50] 李亚舒,黎难秋.中国科学翻译史[M].长沙:湖南教育出版社,2000.

[51] 李银芳.科技英语增译原则[J].中国科技翻译,2007(02):19-21,48.

[52] 梁玉峰.英汉翻译中词义的引申[J].晋中学院学报,2005(01):70-72.

[53] 凌渭民.科技术语翻译法[J].中国翻译,1982(01):42-46.

[54] 凌渭民.科技英语汉译中的词类转换和句子成分转换[J].中国翻译,1982(02):39-43.

[55] 凌渭民.科技英语翻译中的省译法[J].外国语(上海外国语学院学报),1982(01):37-38.

[56] 刘重德.文学翻译十讲[M].北京:中国对外翻译出版公司,1991.

[57] 刘国仕.论词义的理解和选择[J].中国科技翻译,2007(02):15-18.

[58] 刘明东.英语被动语态的语用分析及其翻译[J].中国科技翻译.2001(1):1-4.

[59] 刘庆荣.英语长句的翻译[J].中国翻译,2009,30(06):70-71.

[60] 刘荣廷.科技文体的语言特点和要求[J].科技出版,1989(4):14-17.

[61] 刘向红,罗晓语.科技英语文体的名词化结构及其翻译策略[J].湖南工程学院学报(社会科学版),2015,25(03):43-46.

[62] 刘晓晖.试论文学作品中基本颜色词的翻译方法[J].陕西师范大学学报(哲学社会科学版),2001(S2):311-315.

[63] 刘源甫.科技翻译词义的具体化与抽象化引申[J].中国科技翻译,2004(02):16-19.

[64] 马清海.试论科技翻译的标准和科学术语的翻译原则[J].中国翻译,1997(01):28-29.

[65] 马云.航空器驾驶员低温冰雪运行指南(节选)翻译报告[D].南京:南京航空航天大学,2020.

[66] 马峥嵘,王永胜.谈考研英译汉中非谓语动词的翻译[J].中国科技信息,2007(24):277+280.

[67] 毛荣贵,范武邱.形象思维与科技术语翻译[J].中国科技翻译,2003(04):43-46.

[68] 苗菊,高乾.构建MTI教育特色课程——技术写作的理念与内容[J].中国翻译,2010(2):35-38.

[69] 穆凤良.逻辑比较与英汉翻译[M].北京:国防工业出版社.2009.

[70] 宁玲,赵斌,杨海.航空术语翻译技巧与译句分析[J].中国科技翻译,2012,25(02):5-7.

[71] 牛灵安.科技英语翻译词义的确定[J].中国科技翻译,2004(01):14-16.

[72] 潘文国.汉英语对比纲要[M].北京:北京语言文化大学出版社,1997.

[73] 邵启祥.谈"一词多译"[J].上海科技翻译,1986(03):28-30.

[74] 石春让,胡晓静.英汉科技翻译中的语用失误及规避方略[J].上海理工大学学报(社会科学版),2011,33(03):178-183.

[75] 孙国平.浅谈汉英科技翻译中的词语选择[J].中国科技翻译,1998(02):23-25.

[76] 孙静,王帆,顾飞荣.农业科技论文英译汉中的词类转换[J].中国科技翻译,2014,27(04):4-6+10.

[77] 孙新法.谈科技英语长句的理解与翻译[J].中国科技翻译,2008(04):10-11.

[78] 唐旭光.对英汉翻译引申的反思[J].广西师范学院学报(哲学社会科学版),2009,30(02):108-112.

[79] 田传茂,许明武.试析科技英语中的隐性逻辑关系及其翻译[J].中国翻译,2000(04):56-61.

[80] 田飞龙.词类转译法[J].外语教学,1981(03):51-60.

[81] 田艳.冗余信息与增译和省译[J].中国翻译,2001(05):31-33.

[82] 王爱国.航空英语的构词特点及翻译[J].中国科技翻译,2004(04):8-11.

[83] 王斌.科技英汉翻译中的词义选择[J].华东工业大学学报,1996(02):103-107.

[84] 王博.英语科技术语翻译策略新探[J].华北水利水电学院学报(社科版),2009,25(05):134-136.

[85] 王传英,王斌,张雅雯.技术写作规范研究[J].上海翻译,2016(2):64-70.

[86] 王归立.英汉翻译中的词义引申[J].青年与社会:中外教育研究,2008.

[87] 王家楫.探讨科技翻译中词义的确定[J].中国翻译,1995(03):21-24.

[88] 王金铨.英语定语从句汉译过程的个案研究[J].外语教学与研究,2002(06):471-475.

[89] 王平.科技翻译中的修辞处理[J].中国科技翻译,2011,24(02):5-9.

[90] 汪庆琦.逻辑翻译理论视角下《政策改变促进渔业可持续性与适应性》(节选)英汉翻译实践报告[D].上海:上海海洋大学,2021.

[91] 王泉水.科技翻译中的引申、具体化和概括化处理[J].中国翻译,1990(05):32-36.

[92] 王振平.科普著作的文体与翻译[J].上海翻译,2006(2):35-38.

[93] 王中一,王涯.对增译的几点意见[J].上海科技翻译,1992(04):21-22.

[94]   魏根.篇章语言学与英语语篇的理解[J].河北师范大学学,2006(02):89-92.

[95]   韦孟芬.浅析科技英语翻译的词义选择[J].中国科技翻译,2015,28(01):1-3+35.

[96]   韦孟芬.英语科技术语的词汇特征及翻译[J].中国科技翻译,2014,27(01):5-7+23.

[97]   文军.英汉科技翻译教程[M].上海:上海外语教育出版社,2020.

[98]   文军,李培甲.航空航天科技英语术语翻译研究[J].广东外语外贸大学学报,2011,22(03):27-31+42.

[99]   温雪梅,邱飞燕.科技术语翻译之"约定俗称"与创新[J].中国科技翻译,2010,23(03):13-15+19.

[100]   吴仁华.民航科技英语翻译初探[J].贵州工业大学学报(社会科学版).2003(03):78-82.

[101]   现代汉语词典编写组.现代汉语词典[M].北京:外语教学与研究出版社,2002.

[102]   肖坤学.认知语言学语境下被动句英译汉的原则与方法[J].外语研究,2009(01):17-22.

[103]   谢景芝.英汉被动句的对比与翻译[J].中州学刊,2004(06):200-202.

[104]   谢小红.漫谈科技英语新词及翻译[J].南昌大学学报(人社版).2002(10):145-147.

[105]   徐彬,郭红梅.科普翻译的挑战[J].上海翻译,2012(10):42-46.

[106]   徐莉娜.英译汉中引申依据初探[J].中国翻译,1996(7):35-38.

[107]   许连赞.篇章语言学和阅读教学[J].外国语,1989(04):57-62.

[108]   徐敏娜,骆敏.引申手法在翻译中的应用[J].科技信息(学术研究),2008(13):99-100.

[109]   许云峰.基于信息技术的计算机辅助翻译理论与实践[J].航空科学技术,2014(6):54-57.

[110]   许云峰.英汉航空科技翻译[M].北京:航空工业出版社,2020.

[111]   亚历山大(L.G.Alexander),何其莘.新概念英语第四册[M].北京:外语教学与研究出版社,2000.

[112]   阎德胜.逻辑翻译学构想[J].外语教学,1999(2):42-49.

[113]   严尽忠.英语非谓语动词用法研究[J].教育教学论坛,2017(41):188-190.

[114]   严俊仁.英语被动语态及其应用[M].南京:东南大学出版社,2000.

[115]   杨红.被动语态的翻译研究[J].中国科技翻译,2010,23(03):57-59+23.

[116]   杨兰庆.词义选择与专业知识[J].中国科技翻译,1989(03):25-28.

[117]   杨林.科技英语名物化的语篇衔接功能与翻译[J].中国科技翻译,2013,26(01):1-3.

[118]   杨琦.英译汉翻译技巧之词类转译法[J].西南科技大学学报(哲学社会科学版),2003(04):91-93.

[119]   姚暨荣.论篇章翻译的实质[J].中国翻译,2000(05):20-22.

[120]   叶子南.对翻译中"词性转换"的新认识[J].中国翻译,2007(06):52-53

[121]   余高峰.科技英语长句翻译技巧探析[J].中国科技翻译,2012,25(03):1-3.

[122]   于建平.科技英语长句的分析及翻译[J].中国科技翻译,2000(03):14-16.

[123]   臧金兰.英汉翻译中引申手法的运用[J].山东师大外国语学院学报,2002(04):66-68.

[124]   曾文华,刘萍.类比语料分析对标题与摘要写译的启示——以 Nature 与 Science 原创科研论文为例[J].中国科技期刊研究,2016(2):223-228.

[125]   翟红梅.英汉翻译中的转换手段—引申译法[J].安徽师范大学学报(人文社会科学版),2000(02):264-267.

[126]   翟天利.科技英语突破[M].北京:外文出版社,1999.

[127] 张德富,郭兴家.新概念英语(英汉对照本)第四册[M].合肥:安徽科学技术出版社,1986.

[128] 张辉松.英语名词化结构的语义特征与翻译[J].湖北师范学院学报(哲学社会科学版),2004(02):107-110+112.

[129] 张俊.科技新闻英语的修辞技法及应用初探[J].外语电化教学.2007(117):70-75.

[130] 张梅岗.限制性定语从句传统译法的探讨[J].中国翻译,2000(05):23-26.

[131] 张梅岗,李光曦.科技英语中因果动词的作用与译法[J].中国翻译,1994,(05):21-23+41.

[132] 张培基.英汉翻译教程[M].上海:上海外语教育出版社,2018.

[133] 张珊迪.航空航天科技英语术语翻译分析[J].文化创新比较研究,2018,2(06):84-85.

[134] 张彦.科学术语翻译概论[M].杭州:浙江大学出版社,2008.

[135] 赵红梅.科技英语翻译中词的增译与省译[J].大家,2010(14):105-106.

[136] 赵宏涛.语旨对科技语篇语域特征的影响[J].外语教学,2001(11):21-25.

[137] 赵静.英汉颜色词的比较与翻译[J].解放军外国语学院学报,1999(S1):92-93+110.

[138] 赵萱,郑仰成.科技英语翻译[M].北京:外语教学与研究出版社,2006.

[139] 赵振才,王廷秀.科技英语翻译常见错误分析[M].北京:国防工业出版社,1990.

[140] 郑兰英.翻译中理解障碍问题分析[J].中国科技翻译,2003(02):6-8.

[141] 钟书能.话题链在汉英篇章翻译中的统摄作用[J].外语教学理论与实践,2016(01):85-91+58.

[142] 周方珠.科技翻译的词义选择[J].中国科技翻译,1996(01):2-5.

[143] 周雷敏.国际科技交流中翻译的逻辑判断探析——以材料工程英语为例[J].科技视界.2019:131-132.

[144] 朱波,王伟.论民航专业文本中的术语翻译——以 ICAO 术语为例[J].中国翻译,2013,34(06):94-98.

[145] 朱纯深.浅谈科技翻译中非专业性词语的处理[J].上海科技翻译,1987(03):20-23.

[146] 朱健平.英语长句翻译"五步法"——对比语言学观照下的英语长句翻译思维和教学模式[J].大学教育科学,2007(02):62-65.

[147] 朱俊松.谈科技英语翻译的特点和技巧[J].中国科技翻译,1997(04):13-15.

[148] 朱一凡,胡开宝."被"字句的语义趋向与语义韵——基于翻译与原创新闻语料库的对比研究[J].外国语(上海外国语大学学报),2014,37(01):53-64.

[149] 朱铮铮.对科技翻译特点、标准及要求的探讨[J].科技通报,2009(5):380-383.

[150] 庄一方.专利文献的英汉翻译[M].北京:知识产权出版社,2008.

网络文献

[1] 华盛顿邮报

[2] 环球科学

[3] 经济学人

[4] NASA. The Saturn System:Through the Eyes of Cassini.

[5] https://www.nasa.gov/connect/ebooks/the-saturn-system.html

[6] 纽约时报

[7] 术语在线

［8］ 网易有道词典

［9］ 中华人民共和国科学技术普及法. 北京：中国民主法治出版社，2002.

英文文献

［1］ Halliday，M A K. Cohesion in English［M］. London：Longman Press，1979.

［2］ McEnery A，Xiao Z. Passive Constructions in English and Chinese：a Corpus-based Contrastive Study. Online Proceedings of Corpus Linguistics 2005［EB/OL］. http://www. lancs. ac. uk/postgrad/xiaoz/publications. htm.

［3］ Newmark，P. Approaches to Translation［M］. Oxford：Pergamon，1981.

［4］ Nida E A，Taber C R. The Theory and Practice of Translation［M］. Leiden：E. L. Brill，1982.

［5］ Robinson，D. Western Translation Theory from Herodotus to Nietzsche. Beijing：Foreign Language Teaching and Research Press，2007.